NATURAL HOME CLEANING

OVER 100 WAYS TO CLEAN YOUR HOME NATURALLY

FERN GREEN
PHOTOGRAPHY BY
DEIRDRE ROONEY

Hardie Grant
BOOKS

CONTENTS

INTRODUCTION **06**
Essential Ingredients 08
Essential Oils 10
Using Vinegar 12
Pre-treating Stain Guide 14
Cleaning Kit 16

BATHROOM **19**
All-purpose Bathroom Cleaner 20
Mildew Remover 22
Daily Toilet Cleaner 24
Heavy-duty Toilet Cleaner 26
Citrus & Lavender Toilet Fizz Tablets 28
Lime & Grapefruit Toilet Fizz Tablets 30
Grout Cleaner 32
Whitening Grout Cleaner 34
Daily Bath & Shower Cleaner 36
Intense Bath & Shower Cleaner 38
Bathroom Air Freshener 40
Shower Head Cleaner 42
Bathroom Disinfectant Spray 44
Mirror & Glass Cleaner 46
Lemon & Peppermint Tile Cleaner 48
Shower Curtain Cleaner 50
Rust Remover 52
Air Vent Cleaner 54
Strong Drain Cleaner 56
Bathroom Liquid Hand Soap 58

KITCHEN **61**
Dish Soap 62
Window Cleaner 64
All-purpose Work Surface Spray 66
Stainless Steel Polish 68
Refrigerator Odour Deodoriser 70
Kitchen Bin Deodoriser 72
Dishwasher Detergent (Powder) 74
Dishwasher Detergent (Liquid) 76
Lemon Dishwasher Detergent Tablets 78
Refrigerator & Freezer Cleaner 80
Oven & Hob Cleaner 82
Oven Cleaner 84
Microwave Cleaner 86
Drain Unblocker 88
Chopping Board Cleaner 90
Citrus & Rosemary Liquid Hand Soap 92
Citrus Foaming Hand Soap 94
Antibacterial Handwash 96
Kettle Cleaner 98
Orange & Tea Tree Disinfectant Cleaner 100
Daily Sink Scrub 102
Fruit & Veg Wash 104
Marble Cleaner 106
Granite Cleaner 108
Small Appliance Cleaner 110
Knife Block Cleaner 112
Coffee Pot Cleaner 114
Ice Machine Cleaner 115
Plastic Storage Container Cleaner 116
Silver Cleaner 118
Kids' Cleaning Wipes 120
Hand & Face Wipes 122
Water Purifier 124
Label or Sticky Residue Remover 126

BEDROOM & LIVING ROOM **129**
Air Freshener 130
Air Purifier 132

Damp Defier 133
Carpet Cleaner (Powder) 134
Carpet Stain Remover 136
Hardwood Floor Cleaner 138
Door & Skirting Board Cleaner 140
Wall Scrub 142
Furniture Polish 144
Wood Polish 146
Laminate Floor Cleaner 148
Vinyl Floor Cleaning Spray 150
Tile Floor Cleaner 152
Floor Wipes 154
Wood-cleaning Cloths 156
Disposable Furniture Wipes 158
Scratch Remover 160
Leather Wipes 162
Glass & Mirror Wipes 164
All-purpose Wipes 166
Bed Bug Repellent 168
Antibacterial Wipes 170
Upholstered Fabric Cleaner 172
Linen Spray 174
Quick Insect Repellent for Your Home 176
Moth Repellent 178
Electronic Screen Cleaning Solution 180
Curtain Spray 182
Alternative Insect Repellent Using Dried Herbs 184
Room Scent Spray 186
Smoke Deodoriser 188

UTILITY & LAUNDRY ROOM **191**
Washing Detergent (Powder) 192
Washing Detergent (Liquid) 194
Laundry Tablets 196
Fabric Softener 198
Wet Wipes 200
Shoe Odour Repellent 202
Wine Stain Remover 204
Wrinkle Releaser 206
Washing Machine Cleaner 208
Clothes Whitener 210
Laundry Whitener 212
Stain Spray 214
Tea Tree Stain Stick 216
Oxygen Bleach Paste 218
Linen Freshener 220

GARAGE & SHED **223**
Termite Repellent 224
Ant Repellent 225
Mosquito & Fly Repellent 226
Tick & Flea Repellent for Pets 228
Cockroach Repellent 230
Pet Foot Wipes 232
Garden Tool Cleaning Kit 234
Patio Cleaner 236
Garden Furniture Cleaner 238
Paintbrush Cleaner 240
Car Washing Solution 242
Interior Car Wipes 244
BBQ Grill Cleaner 246
Yoga Mat Cleaner 248
Exercise Machine Spray 250
Trainer Odour Remover 252

Index 254

INTRODUCTION

Are you becoming more conscious about the products you use to clean your home? Are you fed up with using strong chemicals and aware these could be having a detrimental effect on your health? Perhaps you are interested in gaining some knowledge about alternative natural ingredients you could use to refresh and revitalise your home.

In this book, you will find out how to make over 100 products for practically any cleaning dilemma, from natural disinfectants to safe and effective stain removal. Each chapter covers a room in your home, with unique tips and tricks to keep your house sparkling clean, without neglecting the environment or your health.

WHAT ARE THE ADVANTAGES?

There are many advantages in creating your own cleaning products. It's cheaper and safer for the environment and for you and your family. This book will show you how to make everyday products, such as an all-purpose cleaner and laundry detergent, but also the best way to clean things that you don't clean every day, such as curtains or carpet. You can also find out how to scent your homemade products with essential oils to make your home smell how you want it to.

ARE NATURAL CLEANERS EFFECTIVE?

You may think that natural cleaners won't do as good a job as your chemically rich, bleach- and ammonia-filled, shop-bought friends. But you will be pleasantly surprised at how effective these natural recipes are. The idea that you can clean your home and dress your salad with many of the same products might seem a little weird, but once you start creating your own, the results will hugely outweigh any doubts you may have!

HOW SAFE ARE THEY?

Shop-bought cleaning products are composed primarily of water, chemicals and fragrance. Some of these chemicals can aggravate allergies and skin sensitivities as well as pose the threat of poisoning to children and pets. Breathing in the fumes of some cleaners can also pose a health risk. For example, some cleaners instruct you to use a mask or to ventilate a room while using them, but if you need to wear a mask, do you want to be using this in your home? Check the label of your current cleaning products. You will see warning signs, such as 'hazardous to humans and animals' and 'causes substantial but temporary eye injury'. The 'fragrance' used can also be a concern for people who suffer from allergies and asthma.

USING NATURAL CLEANERS

Organic natural cleaners are the only option if you don't want to provoke these sensitivities and allergies. You can also put your mind to rest by knowing exactly what is in them.

BambooDent

ESSENTIAL INGREDIENTS

Many natural cleaners use the same base of ingredients. You will probably find that you have a lot of these in your home already. They are easy to pick up from your local supermarket or online and some are found in the natural beauty section. All products should be stored in a cool, dry place.

DISTILLED WHITE VINEGAR

This is the powerhouse of the cleaning cupboard. It has germ-killing properties, cuts grease and gets rid of stains. It is biodegradable as a mild organic acid and is easy to dispense and control. It is safe for stainless steel, and used by the food industry. It is also relatively non-toxic and stable, so safe for handling, and less likely to leave harmful residues behind. As a sanitiser it is effective against a broad range of bacteria, yeasts and moulds, destroying or reducing these organisms to acceptable levels.

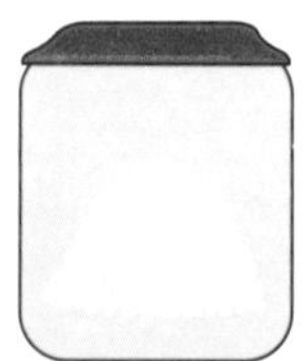

BICARBONATE OF SODA (BAKING SODA)

This is a brilliant product for absorbing odours as well as for its abrasive properties. If you use this a lot, think about buying it in bulk online.

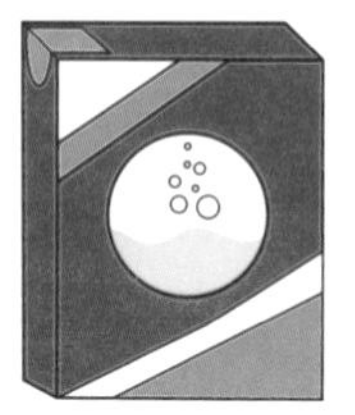

WASHING SODA OR SODIUM CARBONATE

This is an amped-up version of bicarbonate of soda (baking soda) that you mix with water. It is used as a detergent booster and odour and stain remover. Buy it online.

BORAX

This is a naturally occurring mineral found where saltwater lakes and seawater have evaporated. Take care when using this ingredient as there is some controversy over its safety if it is inhaled or ingested. Keep it out of reach of children and pets and thoroughly rinse all surfaces you use it on. Avoid inhaling it when mixing it in a recipe. The results from this product are brilliant.

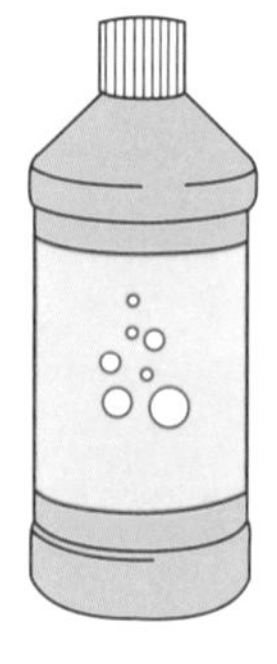

CASTILE SOAP – BAR & LIQUID

This is a vegetable-based soap, making it especially gentle, but the combination of vegetable oils, such as coconut and olive, make it extremely effective in cleaning and cutting down grease. Buy it in most supermarkets or online.

LEMONS & LEMON JUICE

These are great cleaning agents with antibacterial properties and a lovely fresh scent.

SALTS – EPSOM SALTS & SEA SALT

Epsom salts is a compound of magnesium and sulphate and can be ingested for ailments, added to bath water to ease muscle soreness and used to soften water. It works well for cleaning tablets and has some softening properties when used in natural dish soap. Sea salt, however, is a slightly abrasive cleaner. Adding lemon juice to it creates a natural antibacterial scrub.

OILS – ALMOND, OLIVE, VITAMIN E & COCONUT

It is useful to a have a couple of these oils to hand as they can be used to look after wooden surfaces, and in soaps, to nourish the skin.

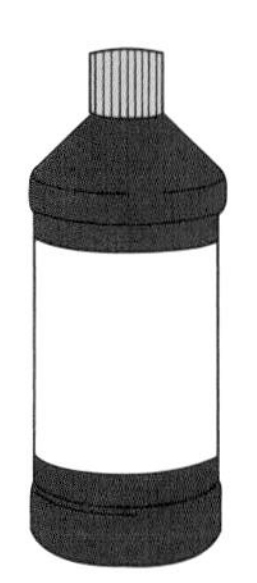

HYDROGEN PEROXIDE

This is found in a bottle in the pharmacy section in the supermarket. It is a natural sanitiser and disinfectant and is so safe you can disinfect toothbrushes with it. It is also unscented and needs to be kept in a dark container. Its scentless and gasless properties are what make it a fantastic disinfectant, and it is great for spraying on wooden boards in the kitchen after you have cleaned them.

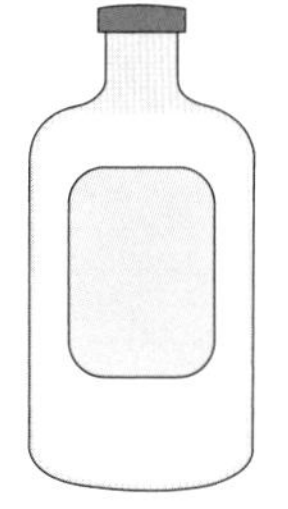

ALCOHOL – RUBBING ALCOHOL & VODKA

Rubbing alcohol, also known as isopropyl alcohol, is found in the pharmacy section in the supermarket, and is sold as an antiseptic. Vodka also kills germs, mould and mildew, and the scent is less noticeable than that of rubbing alcohol.

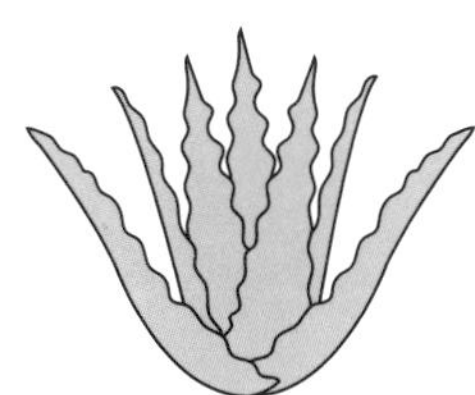

ALOE VERA GEL

Aloe vera is commonly used in beauty care products and cosmetics for its soothing properties. The gel-like substance, which is produced in the plant's leaves, is great for burn relief and use as a hand sanitiser.

ESSENTIAL OILS

Essential oils are vital when making your own products and come in a plethora of fantastic smells. Not only are they high in antibacterial properties but are also antifungal and antiviral, making them an excellent addition to any homemade cleaning recipe. Add oils to your laundry routine, sanitise the kitchen and get rid of mould in the bathroom while making your house smell clean with no scary fumes.

WHAT ARE ESSENTIAL OILS?

Essential oils are not really oils. They do not contain the fatty acids that constitute what we would consider an actual oil. Instead they are highly concentrated plant constituents possessing potent medicinal and cosmetic qualities.

Unfortunately not all essential oils are created equally, so make sure you are purchasing the pure essences to gain maximum benefits from the healthiest and most potent plants. Essential oils are extracted through steam distillation or by cold-pressing the herbs, plants or flowers. That's why the scent is so potent and the beneficial properties are so plentiful and concentrated. All the benefits are distilled into the tiniest droplets and bottled up for your use. Because of the potency, it is recommended to dilute them and not to use them directly on your skin (as they can cause skin reactions or may have other side effects).

PURCHASING ESSENTIAL OILS

When buying essential oils, I recommend visiting a local health food shop and asking what brands they suggest for quality. You can also check out the highly rated ones online. Where possible, try to purchase organic, and check the label to see that the product says 100 per cent pure. That way you know it is not synthetic and that it has not been mixed with anything else. Always check the ingredients list. Don't get essential oils mixed up with scented oils, as these are generally made with synthetic perfumes.

BLENDING ESSENTIAL OILS

When it comes to adding multiple scents to recipes to create a blend, play around with them and start creating mixtures that please your nose.

HERE IS A LIST OF POPULAR BLENDS TO TRY:

Bergamot, lemon & spearmint
Rosemary, lemon & vanilla
Basil, lemon & lime
Lime & lavender
Lavender, rosemary & eucalyptus
Eucalyptus, peppermint & tea tree
Geranium, sandalwood & mandarin
Vanilla & lavender
Vanilla & peppermint
Cinnamon, clove & orange

ESSENTIALS OILS, USES & BLENDS

Here is a table of essential oils with their uses and recommended blends.

ESSENTIAL OIL	USES	BLENDS
Citronella	Insect repellent	Peppermint
Clove	Air freshener	Orange
Eucalyptus	Anti-infective, deodorant	
Grapefruit	Air freshener, antiseptic	Lavender, lemon, lime & orange
Lavender	Antifungal, antiseptic, anti-infective	Bergamot, peppermint, cedarwood, lemongrass, grapefruit, lime, orange, rosemary, ylang-ylang & clary sage
Lemon	Antifungal, antiseptic, anti-infective, antiviral	Basil, bergamot, vanilla, ylang-ylang, rosemary, peppermint, orange & grapefruit
Lime	Air freshener	Orange, grapefruit & lemon
Orange	Air freshener	Lemon, bergamot, grapefruit, lime & clove
Peppermint	Antibacterial, antifungal, antiseptic, antiviral	Lavender & citronella
Rosemary	Antibacterial, antifungal, anti-infective, antimicrobial, antiseptic	Bergamot, thyme, grapefruit, lemon & lavender
Tea tree	Antibacterial, antifungal, antimicrobial, antiseptic, antiviral	Bergamot, clary sage, clove, eucalyptus, lavender, peppermint, geranium & rosemary

USING VINEGAR

Acid cleaners are very versatile when it comes to cleaning, and using a mild acid such as vinegar can be all you need. Vinegar is made from alcoholic fermentation, usually of fruits or grains.

It is comprised of approximately 5 per cent acetic acid and 95 per cent water; however, these proportions can vary. For example, distilled white vinegar usually contains around 5 per cent acidity, whereas Champagne vinegar contains 6 per cent acidity. Vinegars used in household cleaning usually have a 5 per cent acidity level. With its disinfectant properties, due to its pH of 2.0 and acetic acid content, vinegar is an inhospitable environment for many microorganisms.

WHY USE SCENTED VINEGARS?

There is such a wide variety of cleaning uses for vinegar both inside and outside your home, such as getting rid of mildew or cleaning the bath or shower, but perhaps the thought of your home smelling of vinegar is less enticing? Essential oils are a great way to disguise the smell, but you can also mask the smell by making scented vinegars with leftover fruit peels or herbs. Keep your concoctions in the kitchen cupboard, ready to use.

HOW TO MAKE SCENTED VINEGAR

- Wash and dry large glass jars for storing the vinegar. Canning or Kilner (Mason) jars are good, from 500 ml (17 fl oz/2 cups) to 1 litre (34 fl oz/4 cups).
- Fill a jar half full with citrus peels and herbs: lime and mint, orange with sage or orange and basil work well.
- Heat enough white vinegar to fill the jar until it is almost boiling.
- Pour the vinegar over the herbs, seal and place the jars in a cool, dark place.
- After 24 hours, the mix should be ready to use. Allow to stand for longer if you want to deepen the scent.
- Strain the peels and herbs out of the mixture and discard.
- Store in a cool, dark place.

CLEANING IDEAS USING JUST VINEGAR

To get rid of fruit flies – place 80 ml (2½ fl oz/5 tablespoons) vinegar in a small jar and cover with cling film (plastic wrap), securing with an elastic band. Poke a few holes in the cling film with a toothpick. The fruit flies will be attracted to the vinegar and get caught in the jar.

As a rinse aid – place a small dish with 80 ml (2½ fl oz/5 tablespoons) vinegar in the top rack of the dishwasher as a rinse aid, which helps prevent hard water spots and removes detergent residue.

USE SCENTED VINEGARS WITH CITRUS PEEL AND HARDY HERBS TO CLEAN AND FRESHEN UP THE SPACES AROUND YOUR HOME.

PRE-TREATING STAIN GUIDE

It's good to treat stains before washing them. The chart below shows you how to treat everyday stains.

STAIN	TREATMENT
Candle wax	Scrape off surface wax. Place between paper towels and press with a warm iron.
Chocolate	Wet with warm water and dab with Tea Tree Stain Stick (see p. 216).
Fruit	Wet with warm water and dab with white vinegar.
Grass	Pre-treat with Tea Tree Stain Stick (see p. 216). Wash on hottest setting in machine.
Chewing gum	Place an ice cube on the area or put in the freezer. Scrape off and wash as usual.
Lipstick	Dab with rubbing alcohol and soak in Oxygen Bleach Paste (see p. 218). Wash.
Oil & grease	Rub with a piece of chalk to absorb the oil, then wash as usual.
Pen & ink	Rinse with cool water, dab with rubbing alcohol, then wash as usual.
Sweat	Soak in Oxygen Bleach Paste (see p. 218) and warm water, then wash as usual.
Tea & coffee	Use Tea Tree Stain Stick (see p. 216) on area. Wash as usual.
Tomato sauce	Rub a little Castile soap into stain. Leave for a few minutes, then wash as usual.
Vomit & blood	Blot with cold water and add to wash with Oxygen Bleach Paste (see p. 218). Wash.
Water-based paint	Simply rinse using warm water while paint is still wet. Wash as usual.
Wine	Dab with water and soak up stain. Dab with a little vodka or rubbing alcohol and wash as usual.

CLEANING KIT

Before you start mixing up your recipes, it's a good idea to collect together some basic equipment and ingredients that you will frequently use. Some of these you will probably own already and use in various parts of your home, but here are a few items to use as a guide when you start.

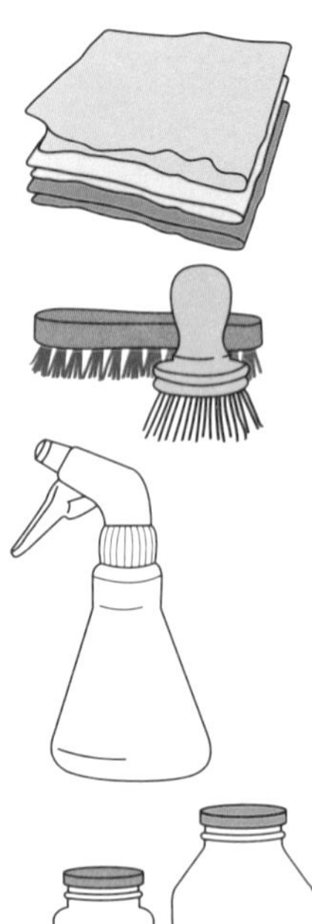

MICROFIBRE CLOTHS

These are made up of a blend of polyester and nylon and are great at not leaving streaks or lint on surfaces. The cloths can be washed many times and reused, making them extremely economical.

SCRUBBING BRUSHES

Use separate scrubbing brushes for specific jobs to avoid cross-contamination.

SPRAY BOTTLES

500 ml (17 fl oz/2 cups) spray bottles will work well for most of the recipes in this book. You can find them online and there are also some glass ones available.

CONTAINERS WITH LIDS

Kilner (Mason) or large jam jars can make attractive storage containers. Don't forget to label them! Keep a stash of jars ready for your own cleaning concoctions.

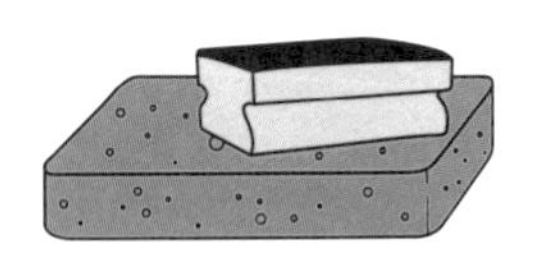

SPONGES

Regular sponges don't contain any odour or cleaner, so they are great to use for your own products. Melamine sponges are particular good for stain removal.

DUSTING WANDS

Find wands with a washable duster and an extendable pole which can reach those tricky corners and tall ceilings.

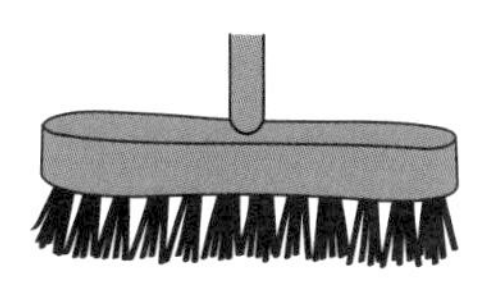

BROOM

A broom and a dustpan are essential for little clean-ups and spills.

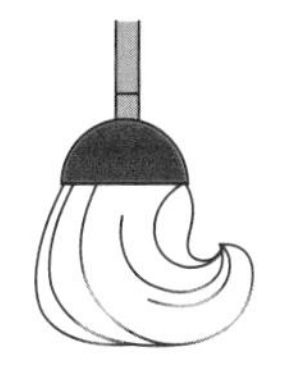

MOP

Find a mop that suits your needs: a sponge mop, a twistable mop or a refillable mop. A refillable mop with microfibre pads can be useful, as these can be washed and used again.

BUCKETS

You need at least three of these: one for a toilet plunger, another for soapy mixtures and another for cleaning rags.

FUNNEL

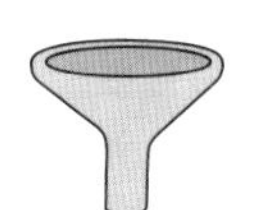

A funnel is very helpful when making up your own recipes. When you have finished mixing a solution, it can help you pour it into a bottle without the solution spilling all over the place.

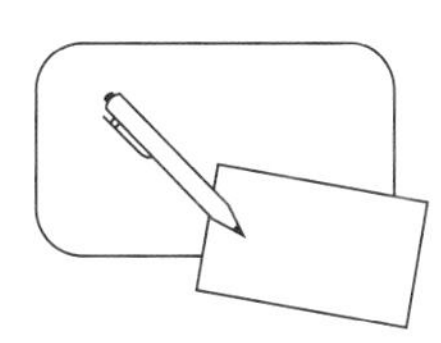

LABELS

An important part of making your own products is labelling them with their contents and the date they were prepared. If you make up the recipe, or adapt one, you could also write the ingredients on the label to remind you of what you used.

CHAPTER 1 BATHROOM

This is a very important room in your home, which gets used several times a day. With so much use, it needs some cleaning attention with good cleaning products. There are many surfaces in the bathroom that will need a regular clean, and this chapter covers them all. Once you discover which recipes work for your bathroom, try keeping the tools that you use to make them in a bathroom cabinet or cupboard to make life easier.

IMPROVE AIR QUALITY

Shop-bought bathroom cleaners can use harmful chemicals, which are not great for the health of your family. Changing to your own recipes and making your own cleaners can not only reduce pollution to our waterways and the air but also minimise the impact on ozone depletion and global climate change. There's no packaging either, so you reduce your waste too!

ALL-PURPOSE BATHROOM CLEANER

S [1] *Soap*	+	H_2O_2 [2] *Hydrogen peroxide*	=	*Cleans & refreshes surfaces*

INGREDIENTS

1 tablespoon hydrogen peroxide
1 teaspoon Castile liquid soap
10 drops peppermint essential oil
5 drops eucalyptus essential oil
5 drops lemon essential oil

PREPARATION 5 minutes | **STORAGE TIME** up to 2 weeks | **STORAGE CONTAINER** small glass or plastic spray bottle

Add the hydrogen peroxide, liquid soap and 2½ tablespoons water to a small spray bottle. Add the essential oils and shake well.

MILDEW REMOVER

INGREDIENTS

100 ml (3½ fl oz/scant ½ cup) white vinegar

2 teaspoons tea tree essential oil

PREPARATION 5 minutes | **STORAGE TIME** 2 months | **STORAGE CONTAINER** 2 glass or plastic spray bottles

Add the white vinegar into one spray bottle and use this first directly on the mouldy areas. If the mould has not gone away after a few hours, mix 500 ml (17 fl oz/2 cups) water and the tea tree essential oil together and decant into the other spray bottle. Shake the bottle well and spray the remaining patches.

DAILY TOILET CLEANER

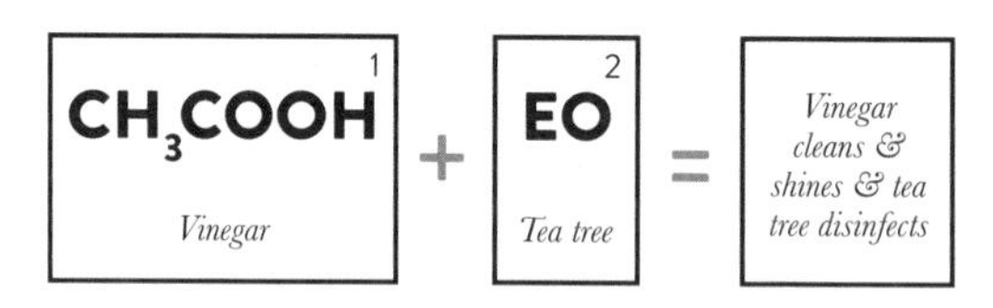

INGREDIENTS

250 ml (8½ fl oz/1 cup) white vinegar

3 drops tea tree essential oil (lemon also works well here)

PREPARATION 5–10 minutes | **STORAGE TIME** up to 2 weeks | **STORAGE CONTAINER** plastic or glass spray bottle

Pour the vinegar into a spray bottle and add a few drops of the essential oil. Shake well and apply the spray to the toilet seat for quick and easy daily cleaning. Wipe with a damp cloth after a few minutes.

This can also be used to keep sinks and bathroom surfaces fresh and germ free. Just spray all over, leave for 5–10 minutes, then wipe with a damp cloth to finish.

HEAVY-DUTY TOILET CLEANER

INGREDIENTS

12 drops tea tree essential oil
60 ml (2 fl oz/¼ cup) white vinegar
125 g (4 oz) bicarbonate of soda (baking soda)

PREPARATION 5–10 minutes | **STORAGE TIME** one-time product, no need for storage
STORAGE CONTAINER none

Put the tea tree oil and vinegar into a bottle, jar or spray bottle. Pour the bicarbonate of soda straight into the toilet bowl and drip or spray the tea tree and vinegar solution into the toilet bowl. Scrub with a toilet brush while the solution fizzes.

CITRUS & LAVENDER TOILET FIZZ TABLETS

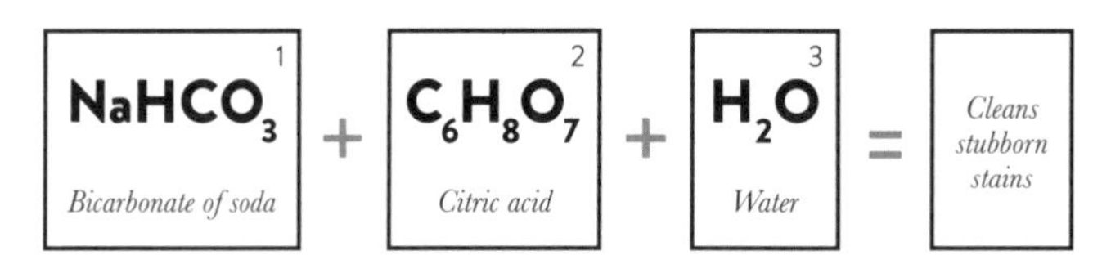

INGREDIENTS

225 g (8 oz) bicarbonate of soda (baking soda)
75 g (2½ oz) citric acid
zest of 1 orange (optional)
25 drops orange or lemon essential oil
24 drops lavender essential oil

PREPARATION 15 minutes, plus overnight | **EXTRA KIT** silicone moulds
STORAGE TIME up to 3 months | **STORAGE CONTAINER** airtight container

Combine the bicarbonate of soda, citric acid and orange zest (if using) in a large bowl. Add the essential oils and stir well to combine. Using a fine mist spray, add 1–2 teaspoons water, ½ teaspoon at a time, and mix with your fingers until it holds together when squeezed but is just damp, not soaked. Press the mixture firmly into silicone moulds, pressing out all the air. Leave overnight before removing. To use, drop one or two tablets into your toilet and leave for 5 minutes. Using a toilet brush, swish your bowl clean.

LIME & GRAPEFRUIT TOILET FIZZ TABLETS

$NaHCO_3$ [1]	+	$C_6H_8O_7$ [2]	+	H_2O [3]	=	Cleans stubborn stains
Bicarbonate of soda		*Citric acid*		*Water*		

INGREDIENTS

225 g (8 oz) bicarbonate of soda (baking soda)
75 g (2½ oz) citric acid
finely grated zest of 1 lime (optional)
25 drops grapefruit essential oil
25 drops lime essential oil

PREPARATION 15 minutes, plus overnight | **EXTRA KIT** silicone moulds
STORAGE TIME up to 3 months | **STORAGE CONTAINER** airtight container

Combine the bicarbonate of soda, citric acid and lime zest in a large bowl. Add the essential oils, stirring well to combine. Using a fine mist spray, add 1–2 teaspoons water, ½ teaspoon at a time, and mix with your fingers until it holds together when squeezed but is just damp, not soaked. Press the mixture firmly into silicone moulds, pressing out all the air. Leave overnight before removing each one. To use, drop one or two tablets into your toilet and leave for 5 minutes. Using a toilet brush, swish your bowl clean.

GROUT CLEANER

INGREDIENTS

125 g (4 oz) bicarbonate of soda (baking soda)
12 drops lemon essential oil
60 ml (2 fl oz/¼ cup) lemon juice
60 ml (2 fl oz/¼ cup) white vinegar

PREPARATION 5–10 minutes | **STORAGE TIME** one-time product, no need for storage
STORAGE CONTAINER none

Fill a 1.5–2 litre (50 fl oz/6 cup–70 fl oz/8 cup) bucket with water. Add all the ingredients and mix well. Use a cloth to apply the solution all along the grout lines and leave it to soak for a few minutes. A hard bristle brush is good for scrubbing the dirt from the grout.

WHITENING GROUT CLEANER

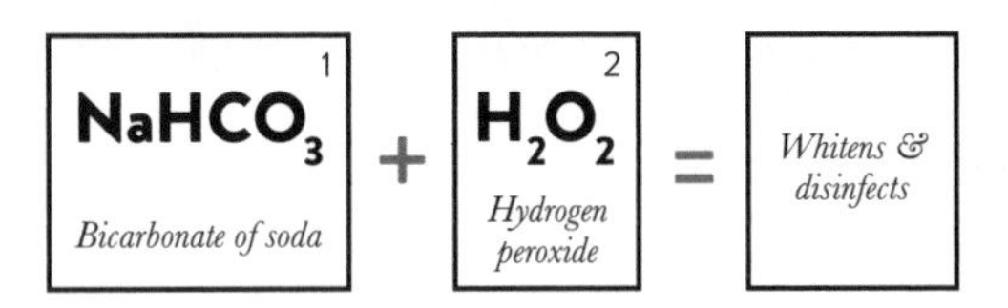

INGREDIENTS

140 g (5 oz) bicarbonate of soda (baking soda)
60 ml (2 fl oz/¼ cup) hydrogen peroxide
juice of ½ lemon

PREPARATION 5–10 minutes | **STORAGE TIME** one-time product, no need for storage
STORAGE CONTAINER none

In a small container or bowl, mix the bicarbonate of soda and hydrogen peroxide together to form a thick paste (fizzing is normal). Stir in the lemon juice to thin the paste. Wet a small cleaning brush, such as an old toothbrush, with water and apply the grout cleaner. Scrub and rinse.

DAILY BATH & SHOWER CLEANER

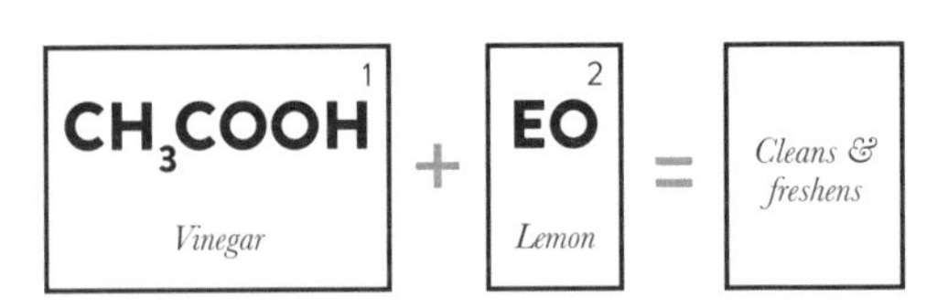

INGREDIENTS

250 ml (8½ fl oz/1 cup) white vinegar

10 drops lemon essential oil

PREPARATION 5 minutes | **STORAGE TIME** up to 3 months | **STORAGE CONTAINER** glass or plastic spray bottle

In a suitable spray bottle, mix 250 ml (8½ fl oz/1 cup) water and the vinegar together, then add the essential oil. Keep it handy for daily cleaning of the bath and shower. Simply spray the surface thoroughly with the mixture, leave to work its magic for a few minutes, then wipe away with a sponge or cloth.

INTENSE BATH & SHOWER CLEANER

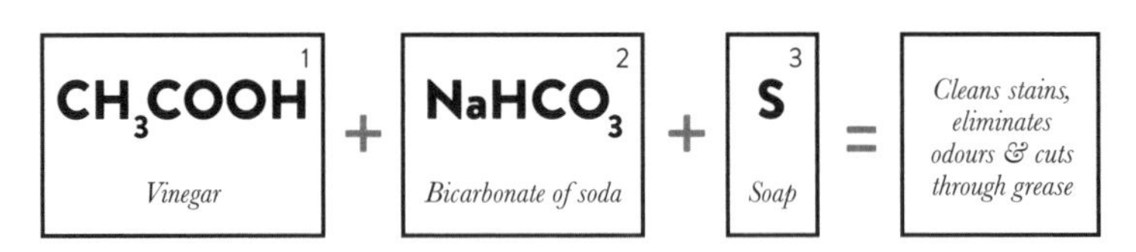

INGREDIENTS

white vinegar, for spraying
2 tablespoons bicarbonate of soda (baking soda)
2 tablespoons Castile liquid soap

PREPARATION 30–40 minutes | **STORAGE TIME** vinegar up to 3 months; bicarbonate of soda and soap – no storage
STORAGE CONTAINER glass or plastic spray bottle

Spray the dirty areas with neat white vinegar and leave it to soak for at least 30 minutes. Meanwhile, mix the bicarbonate of soda and Castile soap together in a separate container. Apply the liquid soap to where the vinegar has soaked and scrub with a hard bristle brush. Rinse with clean water.

BATHROOM AIR FRESHENER

EO[1]	+	EO[2]	=	
Eucalyptus & tea tree		*Lavender & orange*		*Disinfects & freshens*

INGREDIENTS

125 ml (4 fl oz/½ cup) distilled water
6 drops eucalyptus essential oil
6 drops lavender essential oil
6 drops orange essential oil
2 drops tea tree essential oil

PREPARATION 5–10 minutes | **STORAGE TIME** up to 1 month
STORAGE CONTAINER glass or plastic spray bottle

Fill a spray bottle with the distilled water and add the essential oils. Shake well before use.

SHOWER HEAD CLEANER

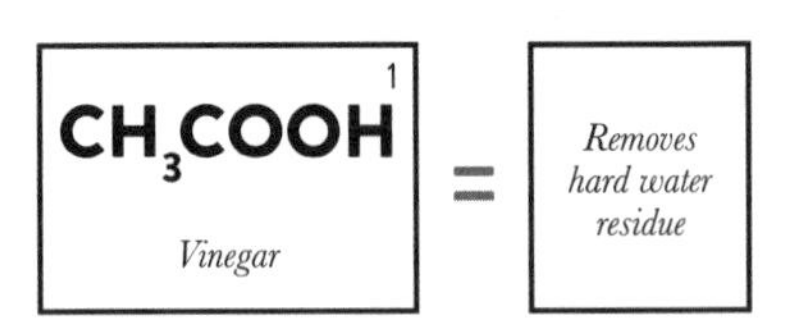

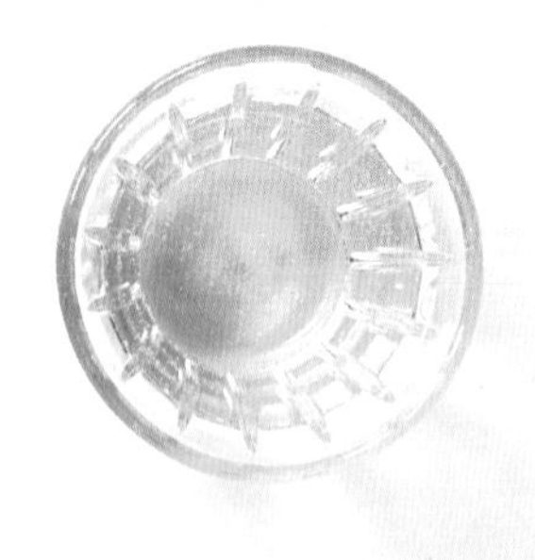

INGREDIENTS

white vinegar

PREPARATION 5 minutes | **EXTRA KIT** heavy-duty plastic bag; elastic (rubber) band
STORAGE TIME 3 months | **STORAGE CONTAINER** none

Fill a heavy-duty plastic bag large enough to hold your shower head with white vinegar. Use an elastic (rubber) band to tie it around the head, making sure all parts of the shower head are submerged in the vinegar. Leave the bag in place for about 10 hours to get rid of any hard-to-reach residue stuck in or on your shower head.

BATHROOM DISINFECTANT SPRAY

S [1] *Soap*	+	EO [2] *Tea tree*	=	*Cleans & disinfects*

INGREDIENTS

3 tablespoons Castile liquid soap

30 drops tea tree essential oil

PREPARATION 5 minutes | **STORAGE TIME** up to 3 months | **STORAGE CONTAINER** glass or plastic spray bottle

Mix 500 ml (17 fl oz/2 cups) water and the soap together, then add the tea tree essential oil. Shake before use.

MIRROR & GLASS CLEANER

CH_3COOH [1]	+	C_2H_5OH [2]	=	*Cleans & evaporates*
Vinegar		*Alcohol*		

INGREDIENTS

60 ml (2 fl oz/¼ cup) rubbing alcohol or surgical spirit
60 ml (2 fl oz/¼ cup) white vinegar
15 g (½ oz/2 tablespoons) cornflour (cornstarch)

PREPARATION 5 minutes | **STORAGE TIME** up to 1 month | **STORAGE CONTAINER** glass or plastic spray bottle

Mix all the ingredients together with 500 ml (17 fl oz/2 cups) warm water, then pour into a spray bottle and off you go. For dirty or streaky mirrors, wipe off with some old newspaper for best results.

LEMON & PEPPERMINT TILE CLEANER

INGREDIENTS

80 ml (2 fl oz/5 tablespoons) white vinegar

1 tablespoon bicarbonate of soda (baking soda)

3 drops peppermint essential oil

5 drops lemon essential oil

PREPARATION 5 minutes | **STORAGE TIME** 2 weeks | **STORAGE CONTAINER** glass or plastic spray bottle

Mix the vinegar, bicarbonate of soda and 360 ml (12 fl oz/½ cup) warm water together in a spray bottle. Top up with the essential oils and shake to combine. Spray on to tiles and leave for up to 15 minutes. Scrub if necessary. Rinse and dry with a clean cloth.

SHOWER CURTAIN CLEANER

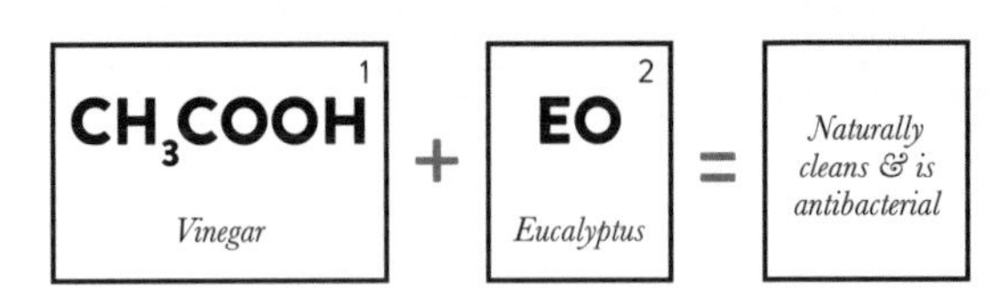

INGREDIENTS

250 ml (8½ fl oz/1 cup) white vinegar
5 drops eucalyptus essential oil
5 drops Dish Soap (see p. 62)

PREPARATION 5 minutes | **STORAGE TIME** up to 2 weeks | **STORAGE CONTAINER** glass or plastic spray bottle

Mix the vinegar with 250 ml (8½ fl oz/1 cup) water in a spray bottle. Top up with the essential oil and washing-up soap drops, and shake to mix. Give your shower curtain a good spray after each use.

RUST REMOVER

INGREDIENTS

white vinegar

PREPARATION 1 minute | **STORAGE TIME** one-time product, no need for storage | **STORAGE CONTAINER** none

Saturate strong paper towels or a cleaning cloth with the vinegar. Leave the cloth draped over the rust stain for 15 minutes, then rub the area with the cloth and the rust will disappear.

AIR VENT CLEANER

CH_3COOH [1] *Vinegar*	+	EO [2] *Lemon & orange*	=	*Cleans & freshens*

INGREDIENTS

500 ml (17 fl oz/2 cups) white vinegar
15 drops lemon essential oil
5 drops orange essential oil

PREPARATION 5 minutes | **STORAGE TIME** up to 2 weeks | **STORAGE CONTAINER** glass or plastic bottle

Pour the vinegar into a glass or plastic bottle and add the essential oils. Dip a cloth in the mix and wring out the excess. Wipe the air vent clean.

STRONG DRAIN CLEANER

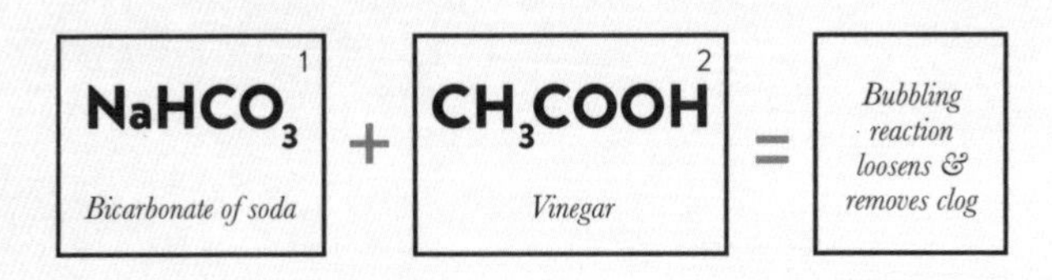

INGREDIENTS

3 drops rosemary essential oil
120 g (4 oz) bicarbonate of soda (baking soda)
120 ml (4 fl oz/½ cup) white vinegar (can use citrus-scented vinegar)

PREPARATION 15 minutes | **STORAGE TIME** one-time product, no need for storage | **STORAGE CONTAINER** none

Put the drops of essential oil down the drain first to add the scent. Sprinkle the bicarbonate of soda down the drain, followed by the vinegar. Leave for up to 15 minutes, then pour a full kettle of boiling water down the drain once the solution has done its magic. Run cold water to clear.

BATHROOM LIQUID HAND SOAP

S [1] *Soap*	+	Vit E [2] *Vitamin E oil*	=	*Cleans & restores*

INGREDIENTS

240 ml (8 fl oz/scant 1 cup) Castile liquid soap
10 drops peppermint essential oil
10 drops lavender essential oil
1 teaspoon vitamin E oil

PREPARATION 5 minutes | **STORAGE TIME** up to 3 months | **STORAGE CONTAINER** soap dispenser bottle

Combine all the ingredients in a glass or plastic soap dispenser bottle. Seal and shake thoroughly to combine.

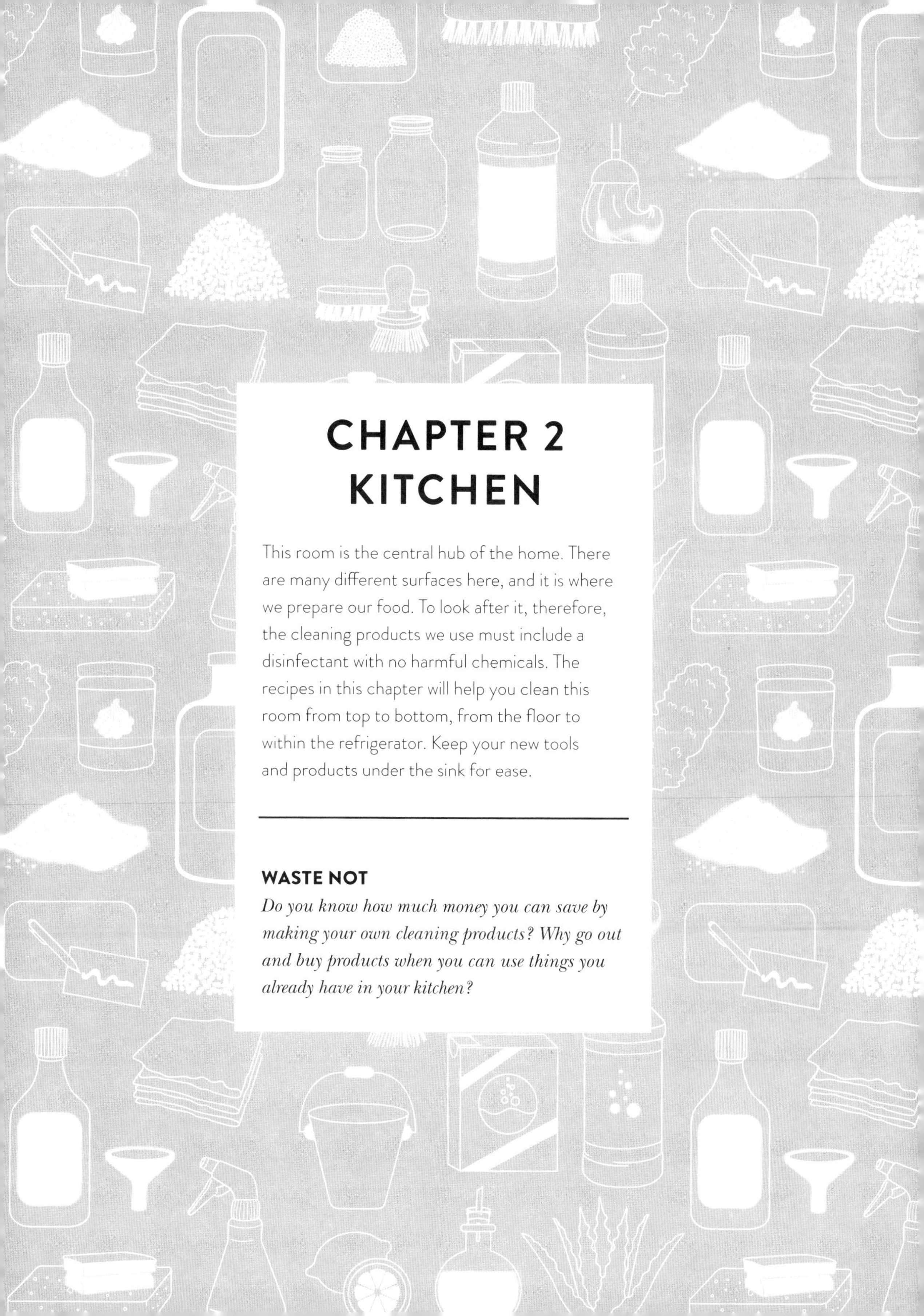

CHAPTER 2
KITCHEN

This room is the central hub of the home. There are many different surfaces here, and it is where we prepare our food. To look after it, therefore, the cleaning products we use must include a disinfectant with no harmful chemicals. The recipes in this chapter will help you clean this room from top to bottom, from the floor to within the refrigerator. Keep your new tools and products under the sink for ease.

WASTE NOT

Do you know how much money you can save by making your own cleaning products? Why go out and buy products when you can use things you already have in your kitchen?

DISH SOAP

S [1] Soap	+	Na_2CO_3 [2] Washing soda	=	Cleans dirt & oil

INGREDIENTS

1½ teaspoons washing soda or sodium carbonate
160 ml (5½ fl oz/⅔ cup) Castile liquid soap
1 teaspoon vegetable glycerine
1 tablespoon aloe vera gel
10 drops lemon essential oil
5 drops peppermint essential oil
5 drops grapefruit essential oil

PREPARATION 15 minutes | **STORAGE TIME** up to 1 month | **STORAGE CONTAINER** soap dispenser bottle

Boil 500 ml (17 fl oz/2 cups) water and leave until warm, then add to the washing soda and stir until dissolved. Add the Castile soap, glycerine, aloe vera and essential oils and stir. Place in the bottle ready to use.

WINDOW CLEANER

L[1] *Lemon* + CH_3COOH[2] *Vinegar* = *Cleans & shines*

INGREDIENTS

juice of ½ lemon

120 ml (4 fl oz/½ cup) white vinegar

1 tablespoon cornflour (cornstarch)

PREPARATION 5 minutes | **STORAGE TIME** up to 1 month | **STORAGE CONTAINER** glass bottle or plastic container

Place the juice, vinegar and cornflour in a bowl and whisk to combine. Add 120 ml (4 fl oz/½ cup) water and fill up a glass bottle or plastic container. Clean the windows with a crumpled-up newspaper dipped in the solution using circular motions.

ALL-PURPOSE WORK SURFACE SPRAY

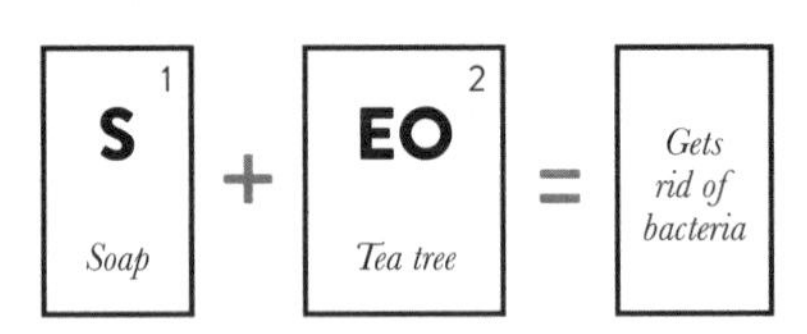

INGREDIENTS

2 tablespoons Castile liquid soap

10 drops tea tree essential oil

PREPARATION 5 minutes | **STORAGE TIME** up to 1 month
STORAGE CONTAINER 500 ml (17 fl oz/2 cup) glass or plastic spray bottle

Add the soap and oil to the spray bottle and top it up with water. Shake to mix.

STAINLESS STEEL POLISH

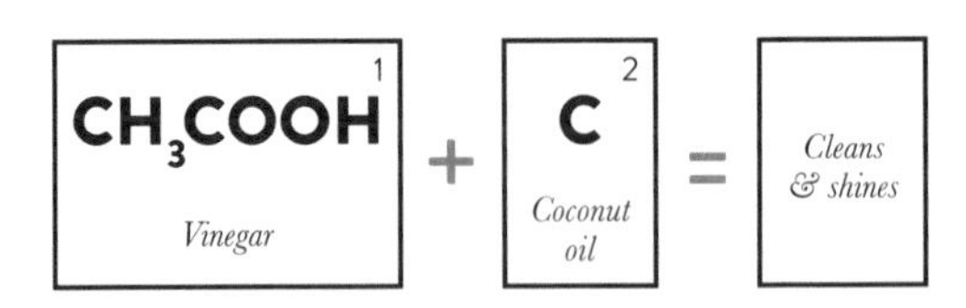

INGREDIENTS

vinegar in a spray bottle (can use a scented vinegar here)
50 ml (1¾ fl oz/3 tablespoons) coconut oil

PREPARATION 5 minutes | **STORAGE TIME** one-time product, no storage necessary | **STORAGE CONTAINER** none

Do a preliminary clean of the stainless steel by spraying the surface with vinegar. Using paper towels or a soft cloth, wipe the vinegar in the direction of the grain. Using a separate clean cloth, apply a small amount of coconut oil directly to the steel and work the oil into the surface. Do this until streaks and stains disappear, then give a last wipe down with an oil-free cloth.

REFRIGERATOR ODOUR DEODORISER

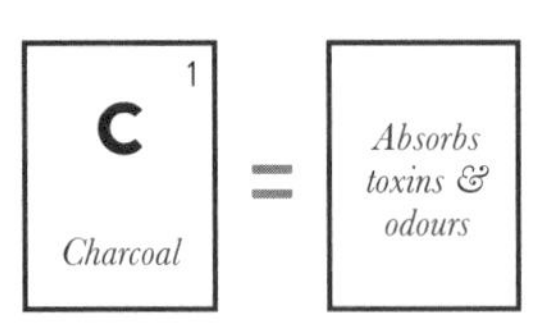

INGREDIENTS

1 binchotan charcoal stick

PREPARATION none | **STORAGE TIME** one-time product, no need for storage | **STORAGE CONTAINER** none

Place the stick in the refrigerator to absorb unwanted odours.

KITCHEN BIN DEODORISER

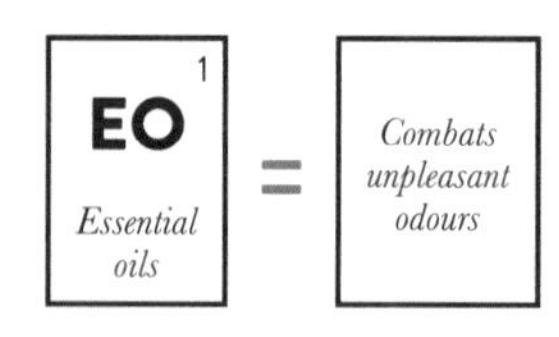

INGREDIENTS

20 drops rosemary essential oil
8 drops grapefruit essential oil
4 drops peppermint essential oil

PREPARATION 5 minutes | **STORAGE TIME** up to 1 year in a cool, dark place
STORAGE CONTAINER small glass bottle

Add all the essential oils to a small glass bottle and shake well to mix. Add a few drops to the bottom of your kitchen bin. This can also be used dropped on to cotton balls and placed in problem areas.

DISHWASHER DETERGENT (POWDER)

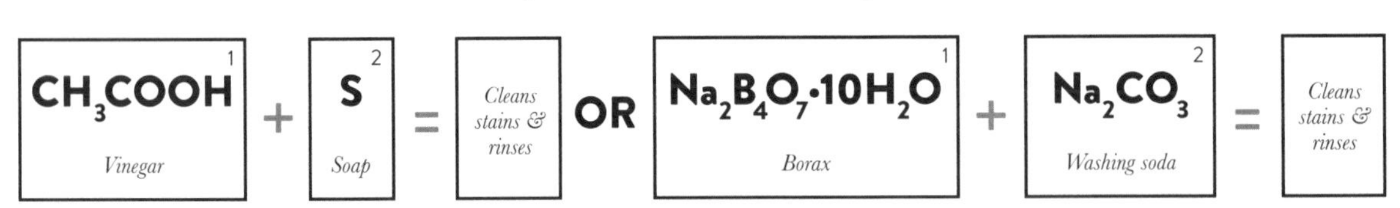

INGREDIENTS

250 ml (8½ fl oz/1 cup) Castile liquid soap
2 teaspoons fresh lemon juice
white vinegar

Or

200 g (7 oz) washing soda or sodium carbonate
200 g (7 oz) borax
100 g (3½ oz) citric acid
100 g (3½ oz) sea or coarse salt

PREPARATION 5 minutes | **STORAGE TIME** 6 months | **STORAGE CONTAINER** any suitable glass or plastic container

Mix the Castile liquid soap, lemon juice and 250 ml (8½ fl oz/1 cup) water together in a container. Add this mixture to one half of your dishwasher's detergent compartment and fill the other half with vinegar. Or, mix the washing soda, borax, citric acid and salt together, then use 1–2 tablespoons per load of dishes.

DISHWASHER DETERGENT (LIQUID)

S[1] *Soap*	+	L[2] *Lemon*	=	*Cleans & shines*

INGREDIENTS

240 ml (8 fl oz/scant 1 cup) Castile liquid soap
1 teaspoon lemon juice
5 drops lemon essential oil
60 ml (2 fl oz/¼ cup) distilled water

PREPARATION 5 minutes | **STORAGE TIME** up to 3 months | **STORAGE CONTAINER** plastic or glass bottle

Mix all the ingredients together and pour into the container. Use 1 tablespoon per load in the dishwasher detergent compartment and run as usual.

LEMON DISHWASHER DETERGENT TABLETS

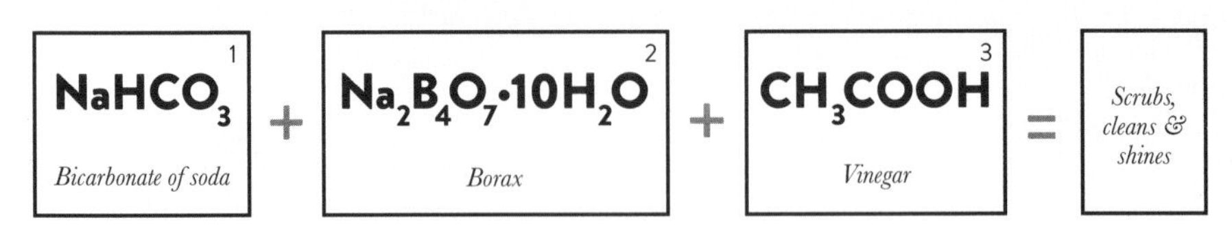

INGREDIENTS

480 g (1 lb 1 oz) bicarbonate of soda (baking soda)
480 g (1 lb 1 oz) borax
120 ml (4 fl oz/½ cup) white vinegar
10 drops lemon essential oil
10 drops peppermint essential oil

PREPARATION 10 minutes, plus overnight | **EXTRA KIT** ice-cube trays
STORAGE TIME up to 2 months | **STORAGE CONTAINER** plastic or glass airtight container

Combine the bicarbonate of soda and borax in a large bowl. Slowly add the vinegar and essential oils and stir to combine. The vinegar will create a fizzing action. When combined, measure a rounded tablespoon and add to an ice-cube tray, pressing down to form a tablet. Allow to dry out and harden overnight. The next day, carefully turn the tray over on to a clean surface. The tablets should fall out. Do not twist or bang as this can cause breakages. Place a tablet in the dish detergent compartment and run as usual.

REFRIGERATOR & FREEZER CLEANER

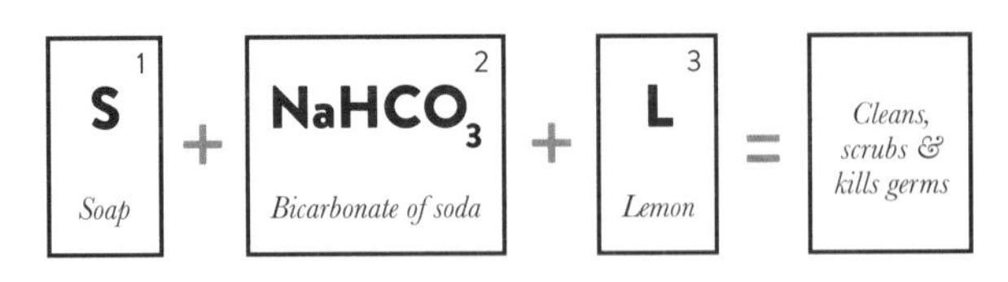

INGREDIENTS

2 tablespoons Castile liquid soap
2 tablespoons bicarbonate of soda (baking soda)
juice from 1 freshly-squeezed lemon

PREPARATION 5 minutes | **STORAGE TIME** one-time product, no need for storage | **STORAGE CONTAINER** none

Combine the Castile soap and bicarbonate of soda in a bucket with 4.5 litres (150 fl oz) hot water. Dip in a cloth and apply to all the nooks and crevices you need cleaned inside the refrigerator. For a fresh touch, go over the inside of your refrigerator with some lemon juice and wipe clean with a cloth.

OVEN & HOB CLEANER

INGREDIENTS

100 ml (3½ fl oz/scant ½ cup) white vinegar

100 g (3½ oz) bicarbonate of soda (baking soda)

PREPARATION 15 minutes | **STORAGE TIME** one-time product, no need for storage | **STORAGE CONTAINER** none

Preheat the oven to 130°C (250°F/gas ½). Using a spray bottle, spray the vinegar generously on areas with stubborn grime. Sprinkle the bicarbonate of soda over the same areas and leave for a few minutes. Turn off the oven and leave to cool. Use a wet cloth or sponge to easily wipe away the mess. Use the same method for the hob, just without the heat.

OVEN CLEANER

INGREDIENTS

50 g (2 oz) coarse salt

150 g (5 oz) bicarbonate of soda (baking soda)

PREPARATION 10 minutes, plus overnight | **STORAGE TIME** one-time product, no need for storage
STORAGE CONTAINER none

To make the paste, combine the salt and bicarbonate of soda with 50 ml (1¾ fl oz/3 tablespoons) water. Before applying, use a sponge to wet the surfaces with water. Spread the paste on the internal parts of the oven. Leave overnight and wipe off with hot water.

MICROWAVE CLEANER

CH_3COOH [1] *Vinegar*	+	L [2] *Lemon*	=	*Cleans & freshens*

INGREDIENTS

100 ml (3½ fl oz/scant ½ cup) white vinegar

2 tablespoons lemon juice

PREPARATION 5 minutes | **STORAGE TIME** one-time product, no need for storage | **STORAGE CONTAINER** none

Fill a glass bowl with vinegar and mix in the lemon juice. Microwave on high for 2 minutes. Leave without opening the microwave door for a few more minutes. Use a cloth to wipe down the inside of the microwave.

DRAIN UNBLOCKER

INGREDIENTS

500 g (1 lb 2 oz) bicarbonate of soda (baking soda) or soda crystals
250 ml (8½ fl oz/1 cup) white vinegar

PREPARATION 10–15 minutes | **STORAGE TIME** one-time product, no need for storage
STORAGE CONTAINER none

Start by pouring 250 g (9 oz) of bicarbonate soda down your drain, followed by 500 ml (17 fl oz/2 cups) boiling water. Leave it a few minutes, then add the remaining 250 g (9 oz) bicarbonate of soda followed by the white vinegar and quickly plug the drain. It should fizz! After a few minutes, flush with some more boiling water.

CHOPPING BOARD CLEANER

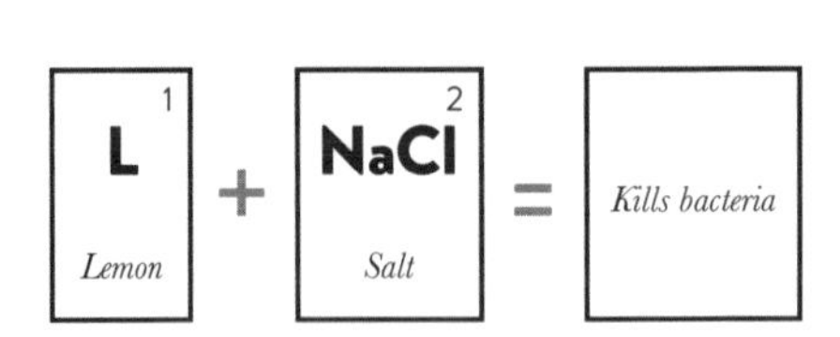

INGREDIENTS

1 fresh lemon

coarse sea salt

PREPARATION 5 minutes | **STORAGE TIME** one-time product, no need for storage | **STORAGE CONTAINER** none

Cut the lemon in half. Sprinkle coarse sea salt all over the chopping board and run a lemon half over the salt, pressing hard to extract as much of the lemon juice as you can. Leave for 10 minutes. Rinse with warm water. Can be used on wooden or plastic chopping boards.

CITRUS & ROSEMARY LIQUID HAND SOAP

S [1] *Soap*	+	Vit E [2] *Vitamin E oil*	=	*Cleans & nourishes*

INGREDIENTS

240 ml (8 fl oz/scant 1 cup) Castile liquid soap
10 drops lemon essential oil
10 drops rosemary essential oil
1 teaspoon vitamin E oil

PREPARATION 5 minutes | **STORAGE TIME** makes 1 soap dispenser | **STORAGE CONTAINER** soap dispenser bottle

Combine all the ingredients in a soap dispenser bottle and shake to mix.

CITRUS FOAMING HAND SOAP

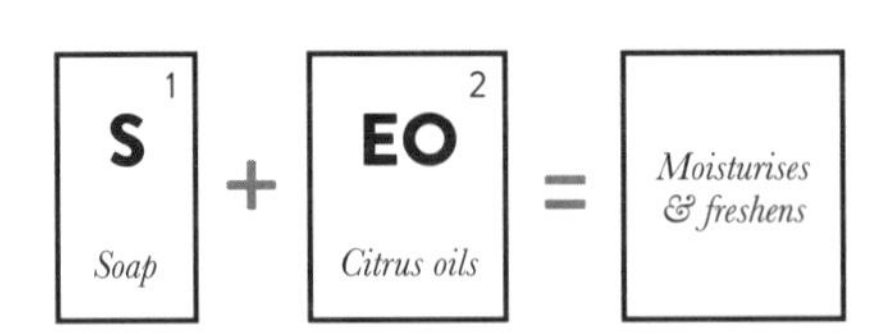

INGREDIENTS

500 ml (17 fl oz/2 cups) distilled water
2 tablespoons Castile soap
1 teaspoon vitamin E oil
10 drops lemon essential oil
10 drops orange essential oil

PREPARATION 5 minutes | **STORAGE TIME** when using distilled water, this can be stored for up to 6 months. If water is not distilled, store for 2 weeks | **STORAGE CONTAINER** soap dispenser bottle

Add the water, soap and vitamin E oil to the dispenser.
Add the essential oils and shake to mix.

ANTIBACTERIAL HANDWASH

C_2H_5OH [1] *Alcohol*	+	EO [2] *Tea tree*	=	*Kills bacteria*

INGREDIENTS

150 ml (5 fl oz/scant ⅔ cup) vodka
10 drops lavender essential oil
30 drops tea tree essential oil
50 ml (1¾ fl oz/3 tablespoons) vitamin E oil

PREPARATION 5 minutes | **STORAGE TIME** up to 3 months
STORAGE CONTAINER small plastic spray bottle or dispenser

Mix all the ingredients well and pour into a spray bottle or dispenser. Apply to your hands to kill bacteria.

KETTLE CLEANER

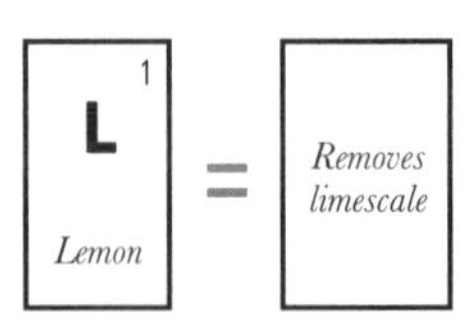

INGREDIENTS

2 lemons

PREPARATION 1 hour | **STORAGE TIME** one-time product, no need for storage | **STORAGE CONTAINER** none

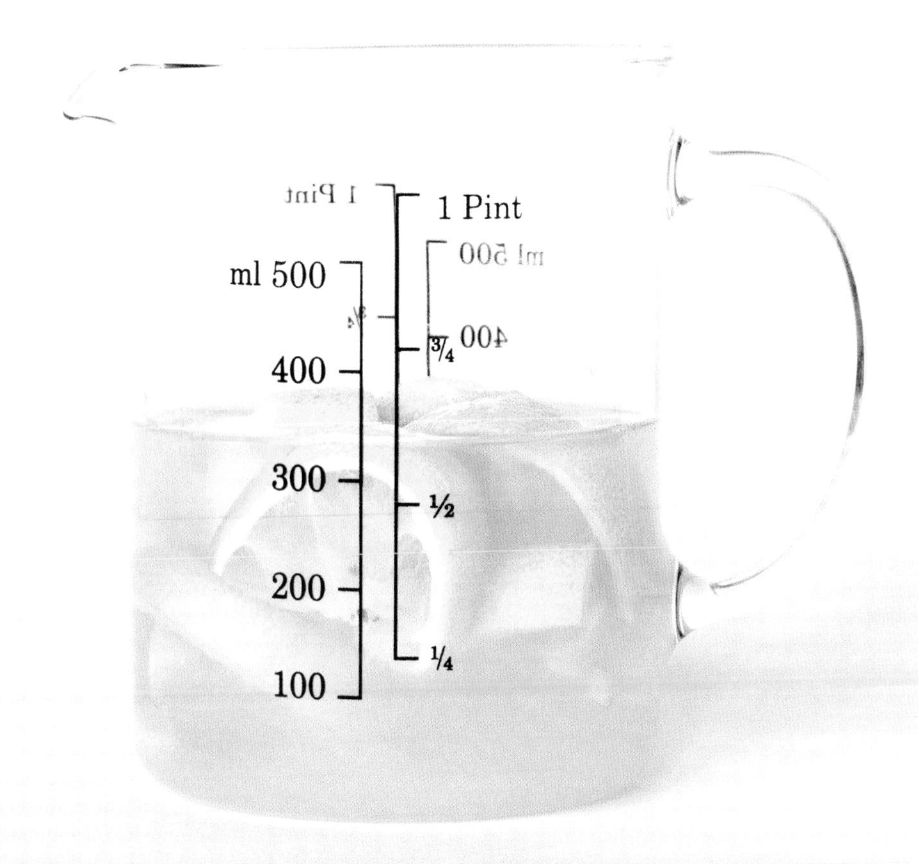

Squeeze the juice of 1 lemon and pour into your kettle. Put slices of the other lemon into the kettle and fill the kettle with water. Put it on to boil. Once boiled, empty all its contents and wash with warm water.

ORANGE & TEA TREE DISINFECTANT CLEANER

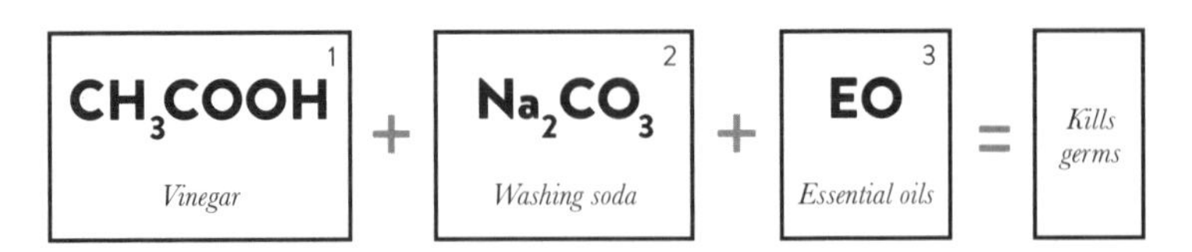

INGREDIENTS

120 ml (4 fl oz/½ cup) white vinegar

1 teaspoon washing soda or sodium carbonate

20 drops tea tree essential oil

10 drops orange essential oil

PREPARATION 15 minutes | **STORAGE TIME** up to 2 weeks | **STORAGE CONTAINER** glass or plastic spray bottle

In a spray bottle, add the vinegar, washing soda and 500 ml (17 fl oz/2 cups) slightly cooled boiled water. Top up with the essential oils and shake to mix. Spray on surfaces and leave for 10 minutes before wiping with a clean cloth or paper towels.

DAILY SINK SCRUB

1 $NaHCO_3$ *Bicarbonate of soda*	+	2 **EO** *Essential oils*	+	3 **S** *Soap*	=	*Cleans grease & grime*

INGREDIENTS

500 g (1 lb 2 oz) bicarbonate of soda (baking soda)
10 drops lemon essential oil
10 drops clove essential oil
1 tablespoon Dish Soap (optional) (see p. 62)

PREPARATION 10 minutes | **STORAGE TIME** indefinite
STORAGE CONTAINER Kilner (Mason) jar or glass container

Add the bicarbonate of soda to a jar or container and scent with the essential oils. Stir to combine. Wet the kitchen sink and sprinkle over 4 tablespoons of the mixture. Add the dish soap to the sink and scrub with a brush. Rinse and air-dry, or dry with a cloth or paper towels.

FRUIT & VEG WASH

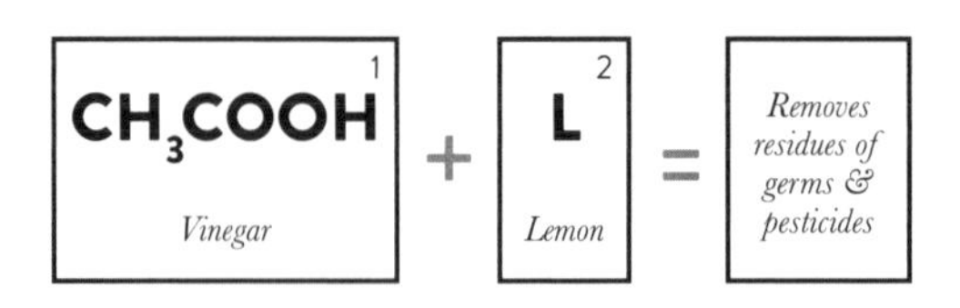

INGREDIENTS

120 ml (4 fl oz/½ cup) apple cider vinegar
120 ml (4 fl oz/½ cup) lemon juice

PREPARATION 5 minutes | **STORAGE TIME** up to 1 month in refrigerator
STORAGE CONTAINER glass or plastic spray bottle

Add all the ingredients, including 120 ml (4 fl oz/½ cup) water, to a spray bottle. Shake to mix, then spray liberally on fruits and vegetables, before rinsing them in cold water and preparing as usual.

MARBLE CLEANER

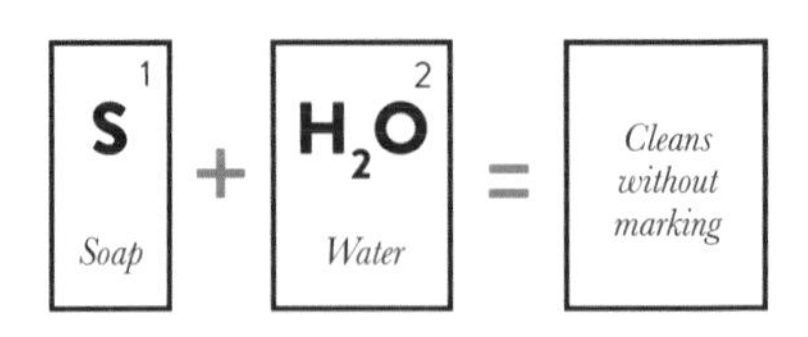

INGREDIENTS

1 teaspoon Castile liquid soap

450 ml (15½ fl oz/1¾ cups) distilled water

PREPARATION 5 minutes | **STORAGE TIME** up to 2 months | **STORAGE CONTAINER** glass or plastic spray bottle

Add the ingredients to a spray bottle and shake to combine. Spray on marble work surfaces and wipe with a damp microfibre cloth or sponge. Dry if necessary.

GRANITE CLEANER

H_2O [1] *Water*	+	C_2H_5OH [2] *Alcohol*	+	S [3] *Soap*	=	*Cleans then evaporates*

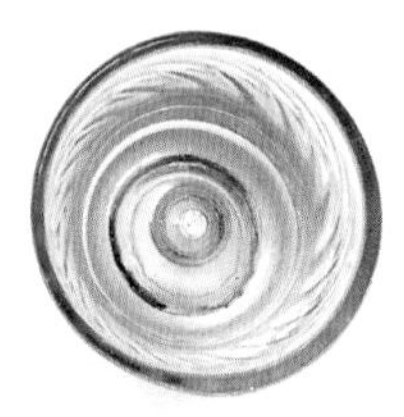

INGREDIENTS

3 tablespoons rubbing alcohol

¼ teaspoon Castile liquid soap

PREPARATION 5 minutes | **STORAGE TIME** up to 1 month | **STORAGE CONTAINER** glass or plastic spray bottle

Add the ingredients, including 450–500 ml (15½ fl oz/1¾ cups–17 fl oz/2 cups) water, to a spray bottle. Shake to combine, then spray on to granite and wipe clean with a microfibre cloth.

SMALL APPLIANCE CLEANER

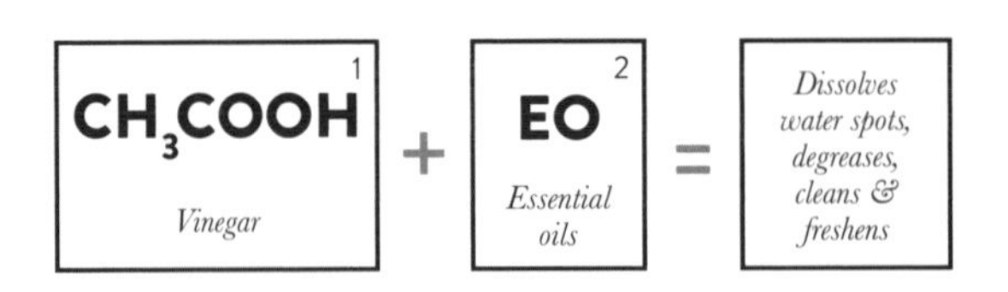

INGREDIENTS

60 ml (2 fl oz/¼ cup) white vinegar

3 drops lemon essential oil

3 drops orange essential oil

PREPARATION 5 minutes | **STORAGE TIME** up to 2 months | **STORAGE CONTAINER** airtight container

Pour the white vinegar into a small container and add the essential oils. Add a little on to a damp cleaning cloth and wipe your small appliance.

KNIFE BLOCK CLEANER

INGREDIENTS

1 teaspoon Castile liquid soap

PREPARATION 5 minutes | **STORAGE TIME** one-time product, no need for storage | **STORAGE CONTAINER** none

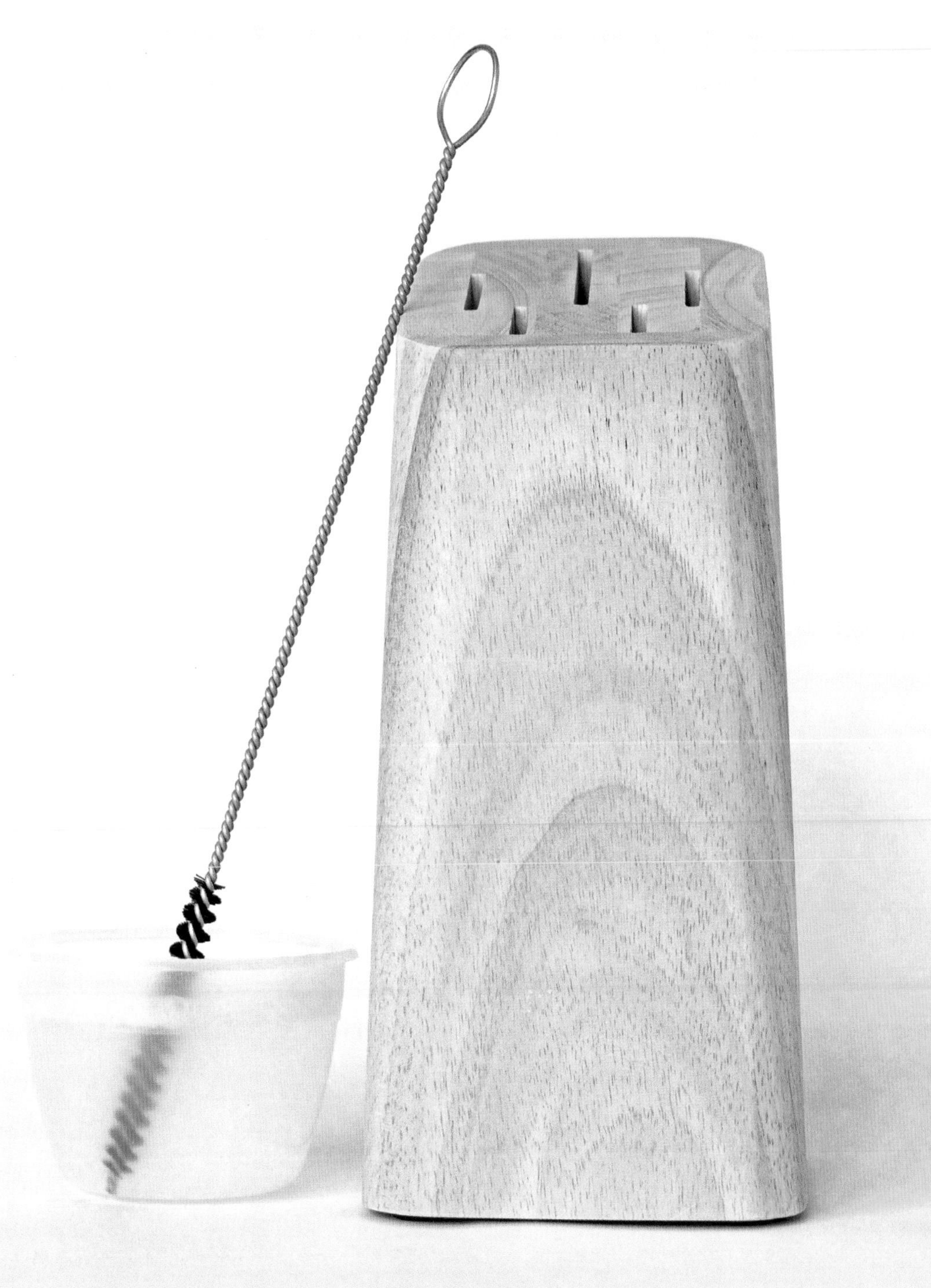

Fill your sink with warm water and add the soap. Dip the knife block in the solution and scrub the holes with a bottlebrush. Rinse thoroughly and allow to air-dry.

PREPARATION 5 minutes | **STORAGE TIME** one-time product, no need for storage | **STORAGE CONTAINER** none

COFFEE POT CLEANER

INGREDIENTS
white vinegar

Fill the coffee pot with equal parts water and vinegar. Run your coffee pot as usual, then empty solution in the sink. Run with two to three cycles of cold water until the vinegar smell is gone and the water runs clear.

PREPARATION 5 minutes | **STORAGE TIME** one-time product, no need for storage | **STORAGE CONTAINER** none

ICE MACHINE CLEANER

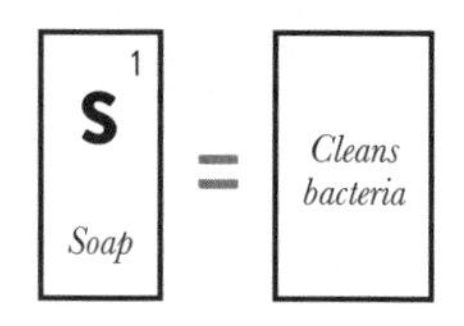

INGREDIENTS

2 drops Castile liquid soap

Add the soap to a bowl of water. Thoroughly wash the ice machine, wiping down the interior. Wipe again with a clean, wet cloth and dry.

PLASTIC STORAGE CONTAINER CLEANER

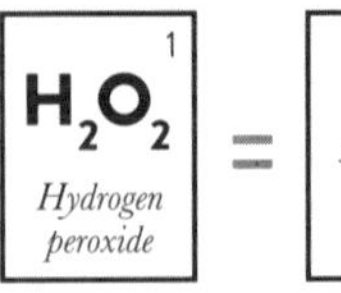

=

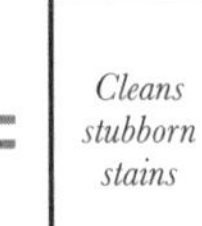

INGREDIENTS

1 tablespoon hydrogen peroxide
1 tablespoon bicarbonate of soda (baking soda)
1 teaspoon Castile liquid soap

PREPARATION 65 minutes | **STORAGE TIME** one-time product, no need for storage | **STORAGE CONTAINER** none

Rinse the plastic container with warm water, then pour the hydrogen peroxide into the bottom. Put the lid on and shake the box so that the peroxide reaches all of the insides. Leave for one hour and then shake again. Sprinkle the bicarbonate of soda into the container and scrub. Wash thoroughly with warm water and soap. Dry with paper towels.

SILVER CLEANER

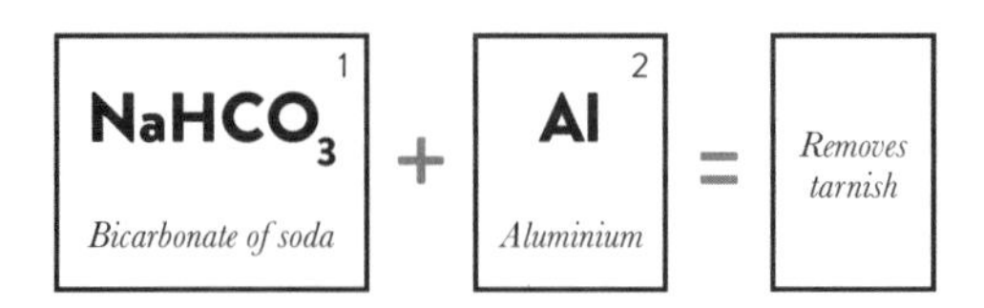

INGREDIENTS

up to 120 g (4 oz) bicarbonate of soda (baking soda), or enough to cover the foil

aluminium foil

PREPARATION 25 minutes | **STORAGE TIME** one-time product, no need for storage | **STORAGE CONTAINER** none

Boil a full kettle. Line a metal tray with foil. Place the silver in the tray and sprinkle over the bicarbonate of soda. Slowly pour boiling water over the silver. Leave for about 15 minutes to remove all tarnish. Wash in soapy water and buff with a soft microfibre cloth.

KIDS' CLEANING WIPES

INGREDIENTS

10–15 washcloths or flannels
240 ml (8 fl oz/scant 1 cup) white vinegar
15 drops tea tree essential oil
15 drops lavender essential oil
5 drops lemon essential oil

PREPARATION 10 minutes | **STORAGE TIME** up to a week | **STORAGE CONTAINER** large airtight glass jar or container

Roll the washcloths and place in a large jar. Combine the vinegar with 250 ml (8½ fl oz/1 cup) water, then top up with essential oils and mix. Pour over the washcloths, pressing down into the liquid. The washcloths should be wet. Depending on the size, you may need to add more vinegar and water as needed in equal parts. Close the lid and use the washcloths as and when you need them.

HAND & FACE WIPES

INGREDIENTS

7–10 small washcloths
120 ml (4 fl oz/½ cup) witch hazel
1 tablespoon olive oil
1 tablespoon Castile liquid soap

PREPARATION 10 minutes | **STORAGE TIME** up to 1 week | **STORAGE CONTAINER** large glass jar or container

Place the small washcloths in a large jar and set aside. Mix 180 ml (6 fl oz/¾ cup) cooled boiled water with the witch hazel, olive oil and Castile soap. Pour the solution over the washcloths and press the washcloths down. Remove a washcloth, wipe your face or hands, rinse in warm water and wipe again to remove any solution.

WATER PURIFIER

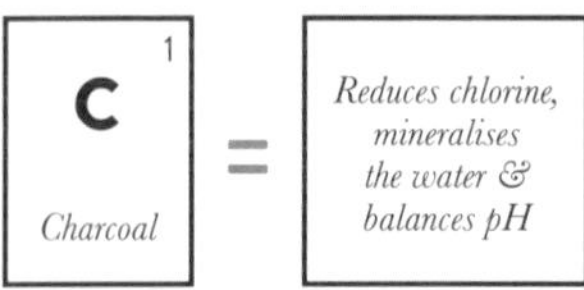

INGREDIENTS

1 binchotan charcoal stick

PREPARATION none | **STORAGE TIME** one-time product, no need for storage | **STORAGE CONTAINER** none

Place the stick in a jug of tap water overnight to purify it. If you are using it once a day, one stick should last 3 months. You can recharge the charcoal once by boiling it for 10 minutes. It should then last another 3 months.

LABEL OR STICKY RESIDUE REMOVER

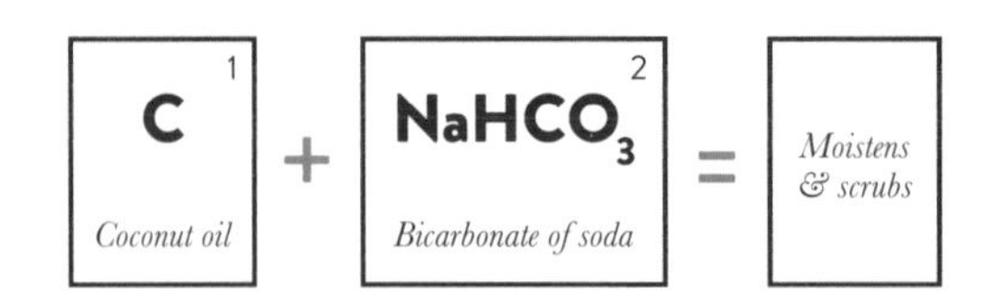

INGREDIENTS

2 tablespoons coconut oil

25 g (1 oz) bicarbonate of soda (baking soda)

PREPARATION 5 minutes | **STORAGE TIME** one-time product, no need for storage | **STORAGE CONTAINER** none

In a small glass or plastic bowl, mix together the ingredients to form a thick paste. Apply directly to the sticky mess and rub with a warm cloth until the residue slips away.

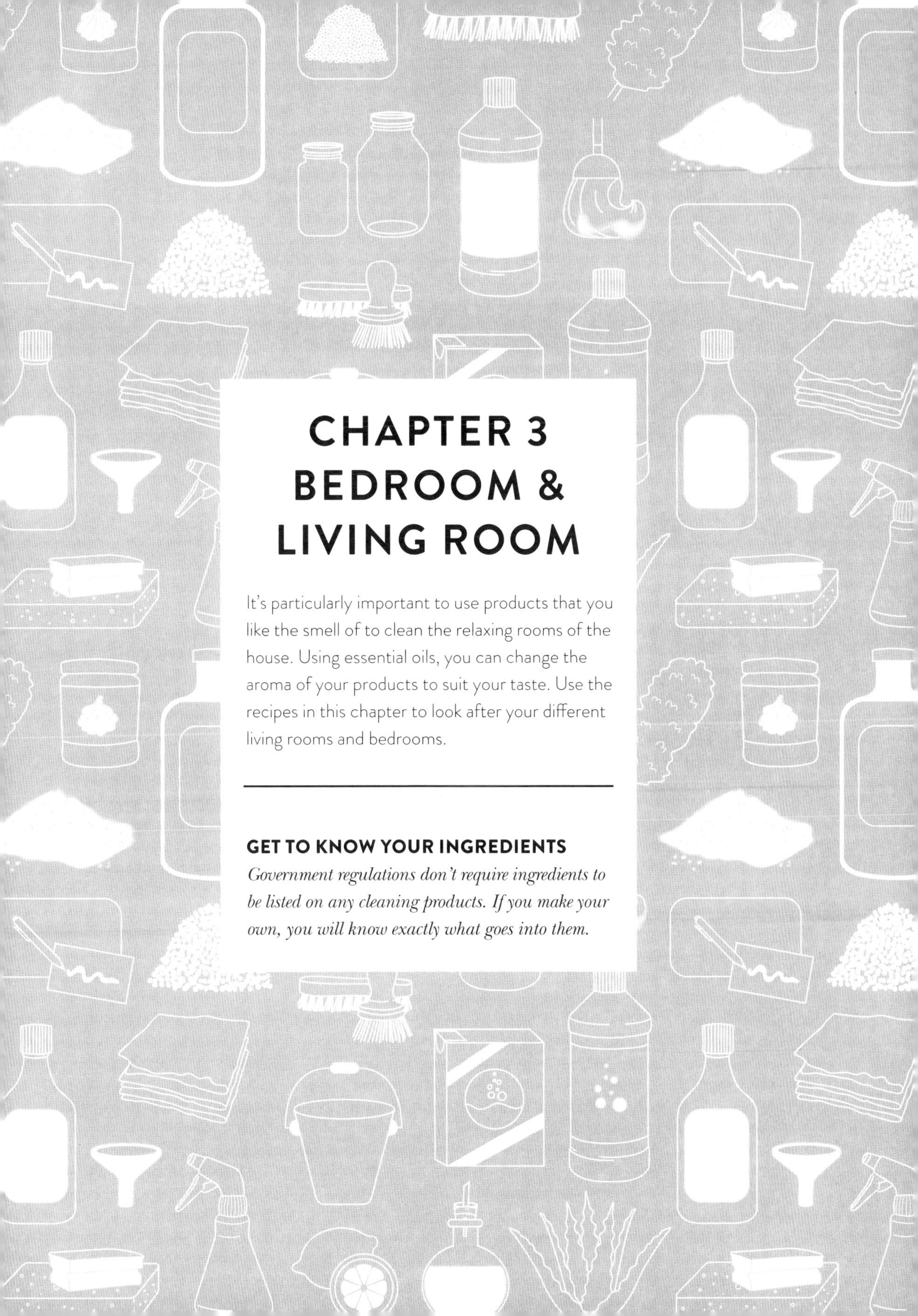

CHAPTER 3
BEDROOM & LIVING ROOM

It's particularly important to use products that you like the smell of to clean the relaxing rooms of the house. Using essential oils, you can change the aroma of your products to suit your taste. Use the recipes in this chapter to look after your different living rooms and bedrooms.

GET TO KNOW YOUR INGREDIENTS

Government regulations don't require ingredients to be listed on any cleaning products. If you make your own, you will know exactly what goes into them.

AIR FRESHENER

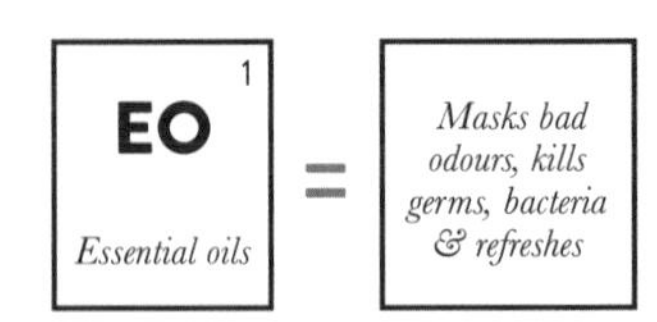

INGREDIENTS

125 ml (4 fl oz/½ cup) distilled water
6 drops eucalyptus essential oil
6 drops lavender essential oil
6 drops orange essential oil
2 drops tea tree essential oil

PREPARATION 10 minutes | **STORAGE TIME** up to 2 months
STORAGE CONTAINER glass or plastic spray bottle, stored in a dark, dry place

Add the water to a spray bottle, then top up with essential oils and shake to combine.

PREPARATION none | **STORAGE TIME** one-time product, no need for storage | **STORAGE CONTAINER** none

AIR PURIFIER

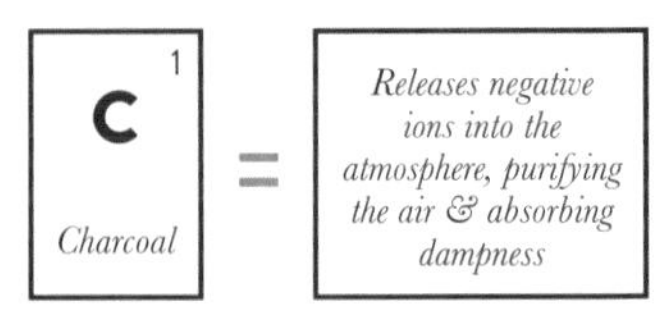

INGREDIENTS

1–2 binchotan charcoal sticks

To purify the air during the night and help you sleep better, place 2 binchotan sticks in a jar in your bedroom.

PREPARATION none | **STORAGE TIME** one-time product, no need for storage | **STORAGE CONTAINER** none

DAMP DEFIER

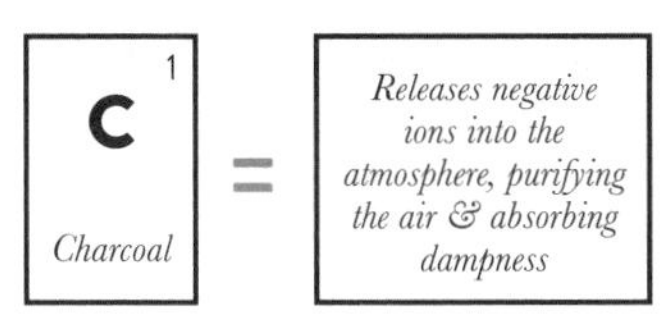

INGREDIENTS

1–2 binchotan charcoal sticks

Place a binchotan stick in a cupboard or drawer to get rid of any damp. Let it dry out in the sun every few weeks.

CARPET CLEANER (POWDER)

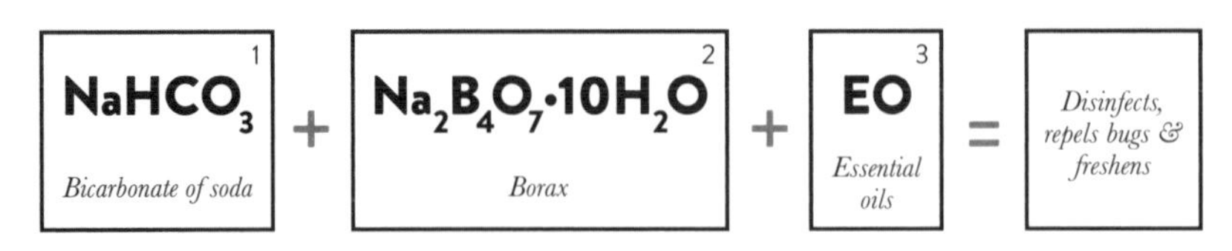

INGREDIENTS

250 g (9 oz) bicarbonate of soda (baking soda)
125 g (4 oz) borax
6 drops lavender essential oil
6 drops tea tree essential oil
6 drops peppermint essential oil

PREPARATION 20 minutes | **STORAGE TIME** up to 2 months | **STORAGE CONTAINER** glass jar

In a glass jar, mix the bicarbonate of soda and borax thoroughly. Drip the essential oils into the mix, shake to combine and sprinkle over your carpet. Leave for 15–20 minutes before vacuuming it all up.

CARPET STAIN REMOVER

INGREDIENTS

500 ml (17 fl oz/2 cups) white vinegar

60 g (2 oz) bicarbonate of soda (baking soda)

PREPARATION 15 minutes | **STORAGE TIME** one-time product, no need for storage
STORAGE CONTAINER glass or plastic spray bottle

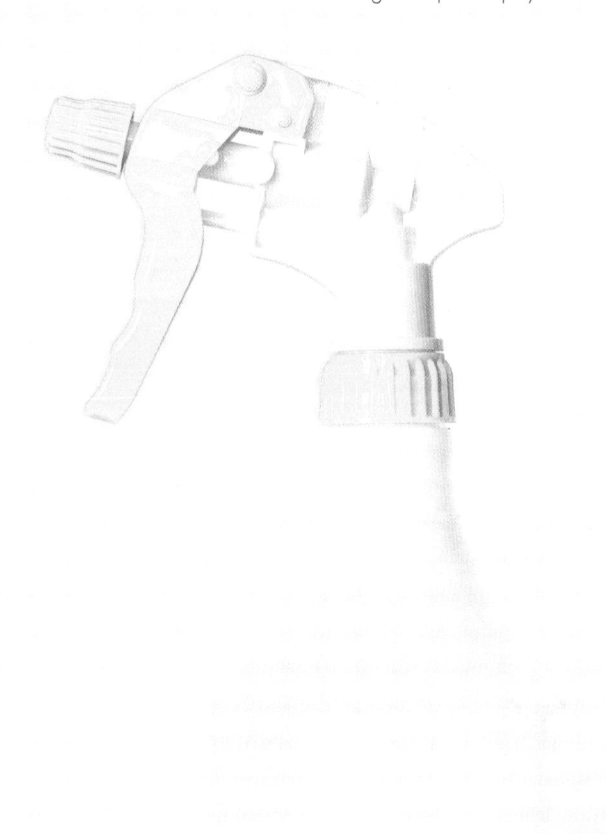

Before applying stain remover, try to remove as much of the stain from the carpet as possible. Using paper towels or a wet cloth, blot up the stain, don't rub. Pour 500 ml (17 fl oz/2 cups) warm water into a large bowl, then add the vinegar, followed by the bicarbonate of soda. (This will fizz.) Pour the mixture into a glass or plastic spray bottle and thoroughly wet the stained area. Leave to soak for 5–10 minutes, then blot off with a damp cloth, then paper towels.

HARDWOOD FLOOR CLEANER

CH_3COOH [1] *Vinegar*	+	EO [2] *Essential oils*	=	*Cleans & freshens*

INGREDIENTS

120 ml (4 fl oz/½ cup) white vinegar
3 drops orange essential oil
2 drops peppermint essential oil

PREPARATION 10 minutes | **STORAGE TIME** one-time product, no need for storage | **STORAGE CONTAINER** none

In a large bucket, mix the vinegar with 5 litres (170 fl oz) warm water. Add the essential oils then, using a mop or by hand with microfibre cloths, clean the floor as usual. Rinse the cloths often as well as the mop, and change the solution when needed.

DOOR & SKIRTING BOARD CLEANER

$Na_2B_4O_7 \cdot 10H_2O$[1] = *Naturally cleans*

Borax

INGREDIENTS

120 g (4 oz) borax

PREPARATION 5 minutes | **STORAGE TIME** one-time product, no need for storage | **STORAGE CONTAINER** none

Stir the borax into 5 litres (170 fl oz) warm water. Dip a microfibre cloth or sponge in the solution, wring out any excess water and wipe doors and skirting boards clean.

WALL SCRUB

INGREDIENTS
60 ml (2 fl oz/¼ cup) white vinegar

PREPARATION 5 minutes | **STORAGE TIME** one-time product, no need for storage | **STORAGE CONTAINER** none

Combine the vinegar with 1 litre (34 fl oz/4 cups) warm water in a bucket. Using a cloth, sponge or brush, scrub the walls clean.

FURNITURE POLISH

L [1] *Lemon*	+	O [2] *Olive oil*	=	*Freshens up wooden furniture*

INGREDIENTS

60 ml (2 fl oz/¼ cup) lemon juice
180 ml (6 fl oz/¾ cup) olive oil
1 drop lemon essential oil

PREPARATION 5 minutes | **STORAGE TIME** up to 1 month | **STORAGE CONTAINER** glass or plastic bottle

Combine the ingredients in a bottle and shake to mix.
Apply sparingly to wooden furniture with a soft cloth.

WOOD POLISH

CH_3COOH [1] *Vinegar*	+	O [2] *Olive oil*	=	*Cleans & restores*

INGREDIENTS

60 ml (2 fl oz/¼ cup) olive oil
60 ml (2 fl oz/¼ cup) white vinegar
10 drops orange essential oil

PREPARATION 5 minutes | **STORAGE TIME** up to 1 month | **STORAGE CONTAINER** glass or plastic bottle

Add all the ingredients to a spray bottle and shake well. Spray on to a cloth (not directly on to wood) and wipe down wooden furniture or other wooden surfaces.

LAMINATE FLOOR CLEANER

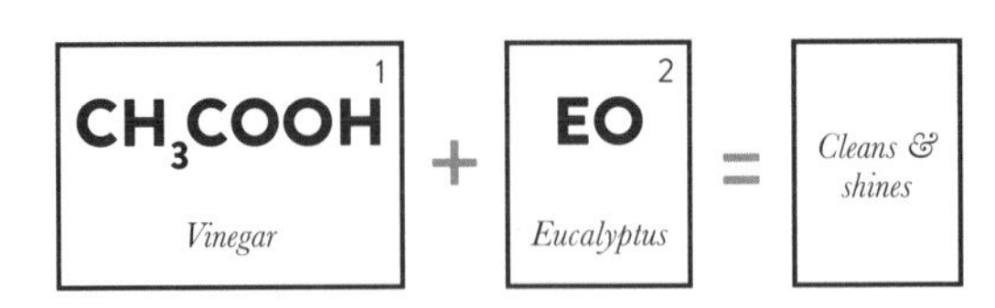

INGREDIENTS

300 ml (10 fl oz/1¼ cups) white vinegar

2 drops eucalyptus essential oil

PREPARATION 5 minutes | **STORAGE TIME** one-time product, no need for storage | **STORAGE CONTAINER** none

In a bucket, mix the vinegar with 300 ml (10 fl oz/1¼ cups) warm water, then add the eucalyptus essential oil. Use a mop or cloth to clean the floor, rinsing regularly.

VINYL FLOOR CLEANING SPRAY

CH_3COOH [1]	+	C_2H_5OH [2]	=	*Cleans, shines then evaporates*
Vinegar		*Alcohol*		

INGREDIENTS

150 ml (5 fl oz/scant ⅔ cup) white vinegar
150 ml (5 fl oz/scant ⅔ cup) rubbing alcohol
150 ml (5 fl oz/scant ⅔ cup) distilled water
3 drops lemon essential oil
2 drops clove essential oil

PREPARATION 5 minutes | **STORAGE TIME** up to 1 month | **STORAGE CONTAINER** glass or plastic spray bottle

Mix the vinegar, alcohol and distilled water in a spray bottle. Add the essential oils and shake. Spray on to the vinyl floor and clean with a wet mop or a wet microfibre mop.

TILE FLOOR CLEANER

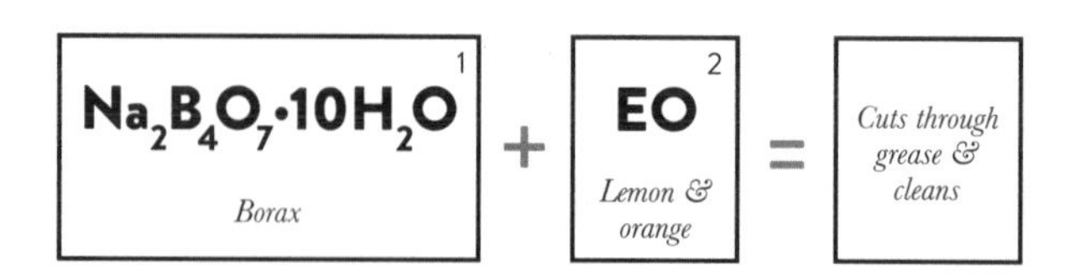

INGREDIENTS

60 g (2 oz) borax
5 drops lemon essential oil
5 drops orange essential oil

PREPARATION 5 minutes | **STORAGE TIME** one-time product, no need for storage | **STORAGE CONTAINER** none

In a mopping bucket, add the borax to 5 litres (170 fl oz) hot water, and top up with the essential oils. Stir to dissolve, then mop to clean the floor; no rinsing necessary.

FLOOR WIPES

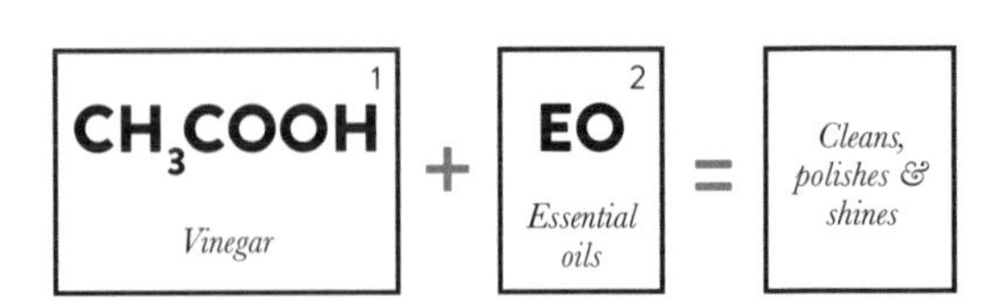

INGREDIENTS

1 litre (34 fl oz/4 cups) distilled water
2 tablespoons white vinegar
3 drops orange essential oil
3 drops tea tree essential oil
8 microfibre cleaning cloths, about 30 x 40 cm (12 x 16 in)

PREPARATION 5 minutes | **STORAGE TIME** up to 1 week

STORAGE CONTAINER airtight container or Kilner (Mason) jar (minimum 1 litre/34 fl oz/4 cups)

Mix the distilled water with the vinegar, then add the essential oils. Place the cloths in the container and pour the solution over them. Close the container and shake it, making sure the cloths have absorbed the liquid. Pour the excess liquid out of the container. Place a cloth over the floor and, using a broom, wipe the floor. One cloth should be enough for a small room. Wash the cloths after use.

WOOD-CLEANING CLOTHS

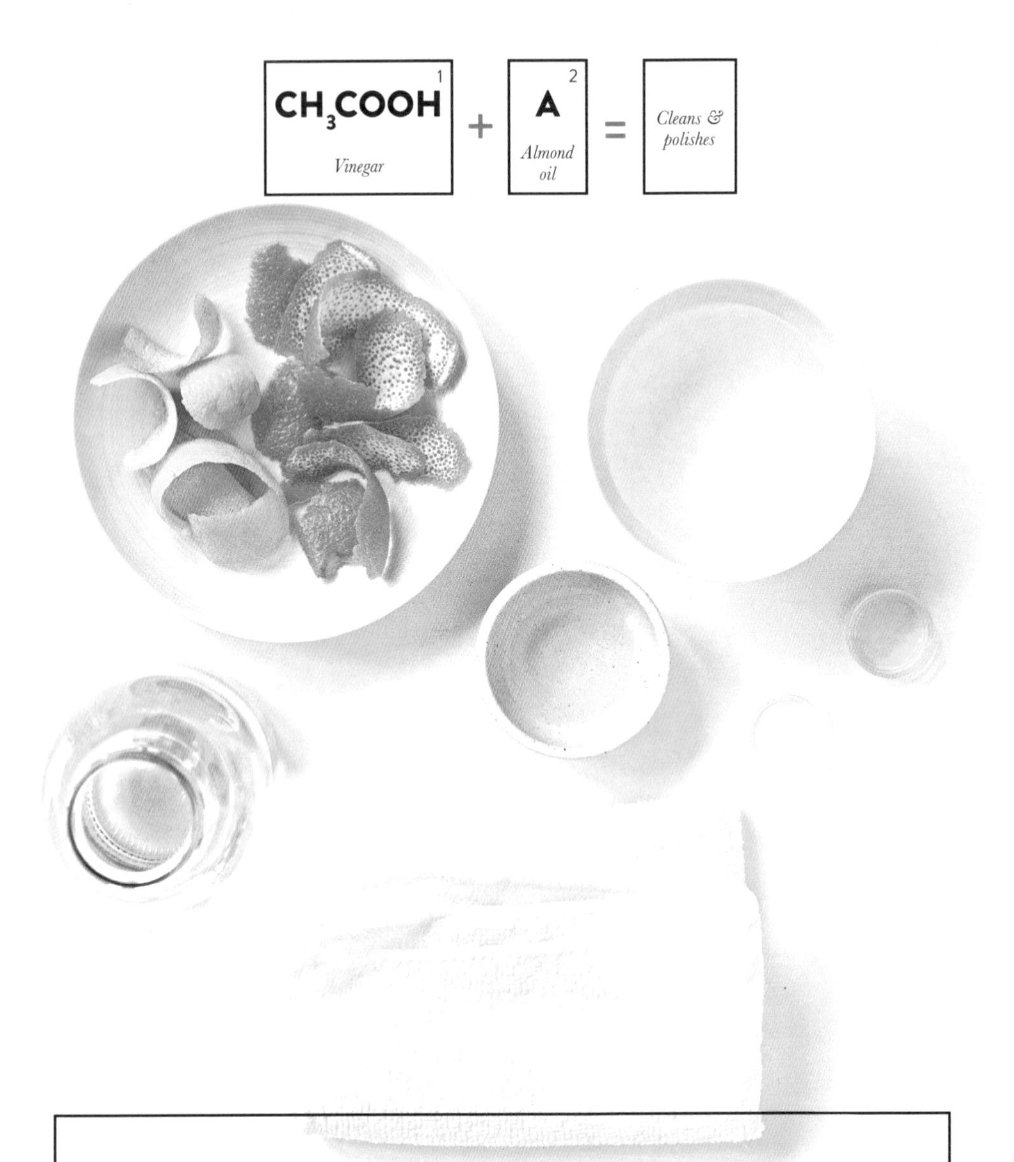

INGREDIENTS

240 ml (8 fl oz/scant 1 cup) white vinegar
500 ml (17 fl oz/2 cups) distilled water
5 drops orange essential oil
5 drops lemon essential oil
1½ tablespoons sweet almond oil
8 microfibre cloths
rind of 1 lemon
rind of 1 orange

PREPARATION 10 minutes | **STORAGE TIME** up to 2 weeks
STORAGE CONTAINER airtight container or Kilner (Mason) jar

Combine the vinegar with the distilled water, then top up with the essential oils and almond oil. Place the cloths in the container and add the lemon and orange rinds. Swirl the liquid around the container and pour off the excess, squeezing cloths to get the liquid out. Cloths should be barely damp. Use to wipe down wooden furniture as needed.

DISPOSABLE FURNITURE WIPES

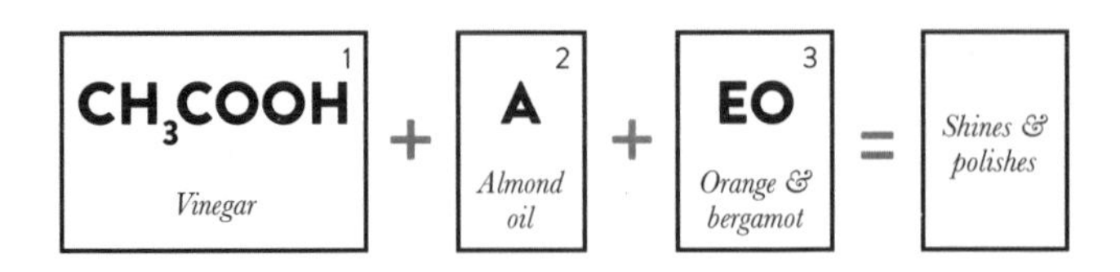

INGREDIENTS

12 sheets thick paper towel
60 ml (2 fl oz/¼ cup) white vinegar
5 drops orange essential oil
5 drops bergamot essential oil
1 tablespoon sweet almond oil

PREPARATION 5 minutes | **STORAGE TIME** up to 1 week | **STORAGE CONTAINER** airtight container

Cut the paper towels in half and stack in a pile. Roll them up and stuff them in the container. (You can also use a ziplock bag.) Mix the vinegar with 250 ml (8½ fl oz/1 cup) water and add the essential oils. Pour over the paper towels and cover with a lid. Shake well. Carefully remove a paper towel or two and wipe furniture. Buff with the almond oil using a microfibre cleaning cloth.

SCRATCH REMOVER

INGREDIENTS

1 teaspoon lemon juice

1 teaspoon white vinegar

PREPARATION 5 minutes | **STORAGE TIME** one-time product, no need for storage | **STORAGE CONTAINER** none

Mix the ingredients together in a bowl. Dip a cloth into the solution and rub into scratches until they disappear. Buff away any residue with a dry cloth.

LEATHER WIPES

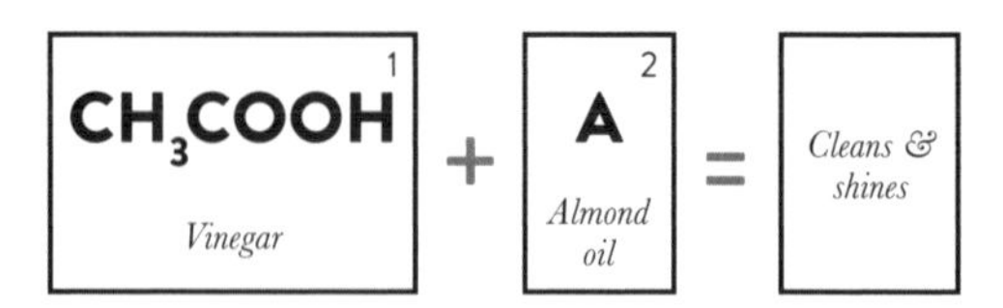

INGREDIENTS

12 sheets strong paper towel
120 ml (4 fl oz/¼ cup) white vinegar
3 tablespoons almond oil
120 ml (4 fl oz/¼ cup) distilled water

PREPARATION 10 minutes | **STORAGE TIME** up to 1 week

STORAGE CONTAINER airtight container, Kilner (Mason) jar or ziplock bag

Cut the paper towels in half and leave them in a pile. Roll up the towels and stuff them in your container. Mix the ingredients with the distilled water, then pour over the towels, seal, swirl and shake. Pour off any excess liquid as wipes need to be on the dry side. Remove a wipe or two, wring out the excess cleaner into your sink, and use as you would any cleaning wipe.

GLASS & MIRROR WIPES

INGREDIENTS

12 sheets strong paper towel
2 tablespoons rubbing alcohol
1 tablespoon white vinegar
5 drops peppermint essential oil
240 ml (8 fl oz/scant 1 cup) distilled water

PREPARATION 10 minutes | **STORAGE TIME** up to 1 week
STORAGE CONTAINER airtight container, Kilner (Mason) jar or ziplock bag

Cut the paper towels in half and stack them in a pile. Roll up the towels and stuff them in your container. Mix the ingredients with the distilled water in a jug and pour over the towels in the container. Cover with a lid or ziplock and shake the liquid around. Pour off the excess liquid as the wipes work better if they are a little on the dry side. To use, carefully remove a wipe, wring out any excess solution into your sink and use.

ALL-PURPOSE WIPES

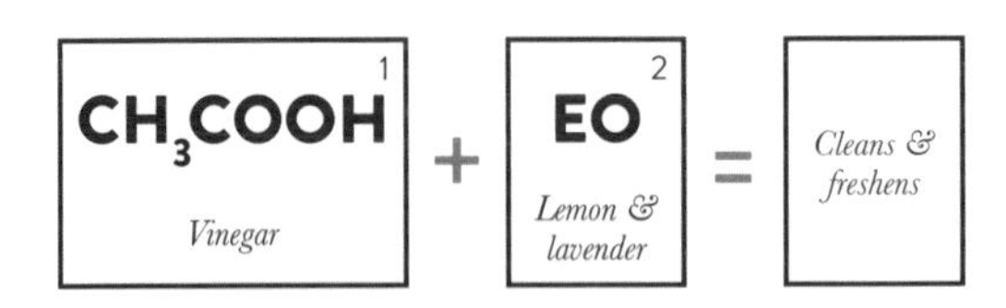

INGREDIENTS

60 ml (2 fl oz/¼ cup) white vinegar
5 drops lemon essential oil
5 drops lavender essential oil
500 ml (17 fl oz/2 cups) distilled water
12 sheets strong paper towel

PREPARATION 5 minutes | **STORAGE TIME** up to 1 week
STORAGE CONTAINER airtight container, Kilner (Mason) jar or ziplock bag

Mix the vinegar and oils with the distilled water. Cut the paper towels in half and stuff them into your container. Pour over the solution and swirl and shake it around until the solution is absorbed. Empty any excess liquid. When ready to clean, remove a paper towel carefully, wring out any excess into the sink and use.

BED BUG REPELLENT

G [1]	+	EO [2]	=	*Repels bed bugs*
Garlic		*Essential oils*		

INGREDIENTS

250 ml (8½ fl oz/1 cup) distilled water
2 garlic cloves, peeled
5 drops rose geranium essential oil
5 drops rosemary essential oil
5 drops tea tree essential oil

PREPARATION 10 minutes | **STORAGE TIME** up to 1 month | **STORAGE CONTAINER** small glass or plastic spray bottle

Fill the spray bottle with the distilled water and add the garlic and essential oils. Shake and leave for a few minutes before use. Shake well again and apply a fine mist to the mattress. Repeat for 3 consecutive days.

ANTIBACTERIAL WIPES

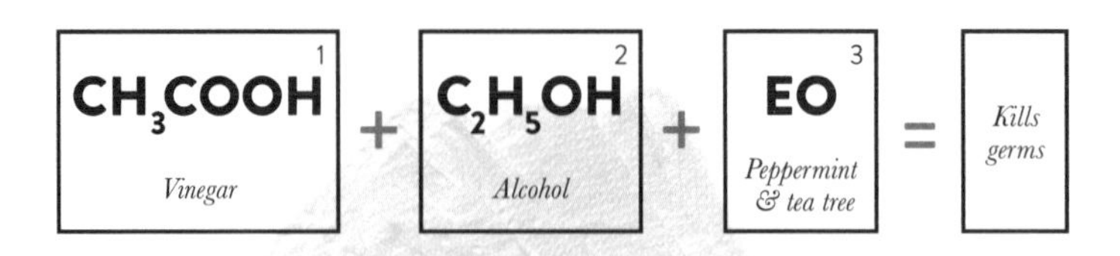

INGREDIENTS

60 ml (2 fl oz/¼ cup) white vinegar
60 ml (2 fl oz/¼ cup) vodka
5 drops peppermint essential oil
5 drops tea tree essential oil
12 sheets strong paper towel

PREPARATION 10 minutes | **STORAGE TIME** up to 1 week
STORAGE CONTAINER airtight container, Kilner (Mason) jar or ziplock bag

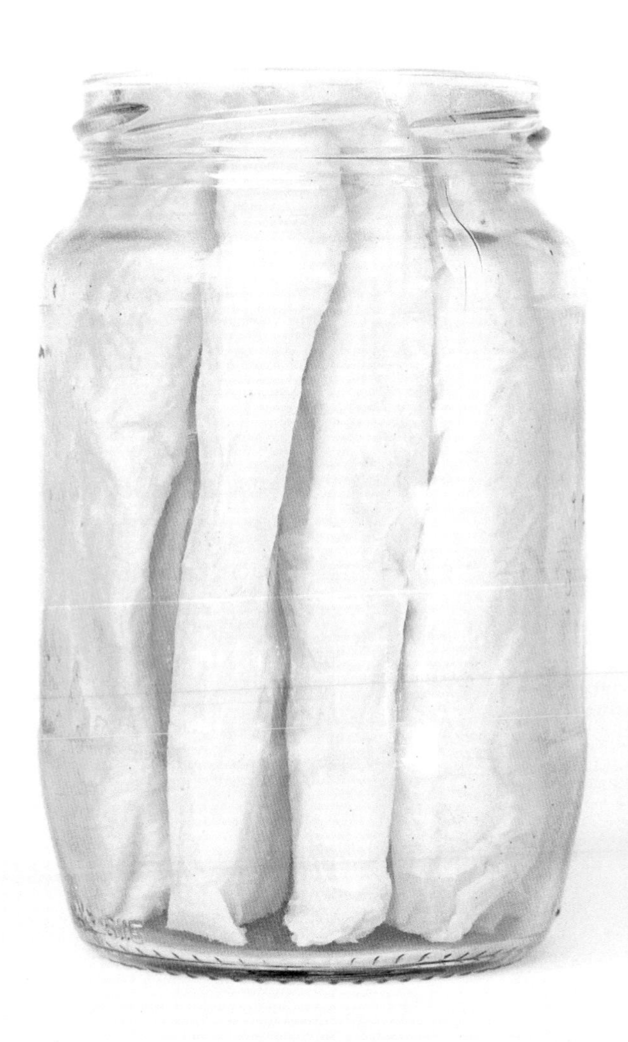

Cut the paper towels in half, roll them up and put them in your container. Mix all the ingredients together in a jug and pour them over the towels. Seal the container and shake so the sheets have absorbed as much as they can. Pour out any excess liquid and seal. When ready to use, carefully remove a towel, wipe the surface and let it remain wet for 5–10 minutes for the solution to work.

UPHOLSTERED FABRIC CLEANER

S [1] *Soap*	+	H_2O [2] *Water*	=	*Removes stains*

INGREDIENTS

2 teaspoons Castile liquid soap

PREPARATION 5 minutes | **STORAGE TIME** one-time product, no need for storage | **STORAGE CONTAINER** none

Mix the soap with 60 ml (2 fl oz/¼ cup) warm water in a bowl. Dip a cotton cloth in the solution and dab away at the stain. When the stain is removed, dab it with a dry cloth to absorb any liquid.

LINEN SPRAY

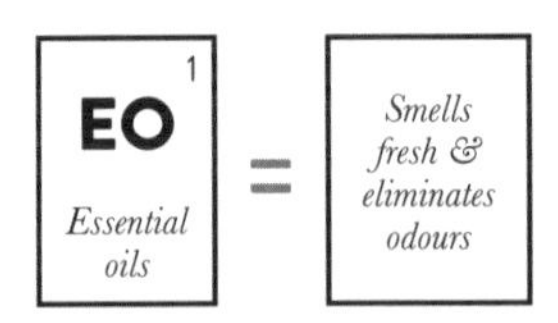

INGREDIENTS

7 drops lavender essential oil
5 drops cedarwood essential oil
3 drops eucalyptus essential oil
250 ml (8½ fl oz/1 cup) distilled water

PREPARATION 5 minutes | **STORAGE TIME** up to 3 months in a cool, dark place
STORAGE CONTAINER small glass or plastic spray bottle

Add the ingredients to a small spray bottle and shake well. Use as a spray or dot the mixture on cotton balls and tuck them into the linen closet.

QUICK INSECT REPELLENT FOR YOUR HOME

W [1] *Witch hazel*	+	EO [2] *Essential oils*	=	*Powerful repellent*

INGREDIENTS

100 ml (3½ fl oz/scant ½ cup) witch hazel
20 drops clove essential oil
10 drops rosemary essential oil
15 drops lavender essential oil
100 ml (3½ fl oz/scant ½ cup) distilled water

PREPARATION 5 minutes | **STORAGE TIME** up to 1 month
STORAGE CONTAINER minimum 200 ml (7 fl oz/1 cup) small glass spray bottle

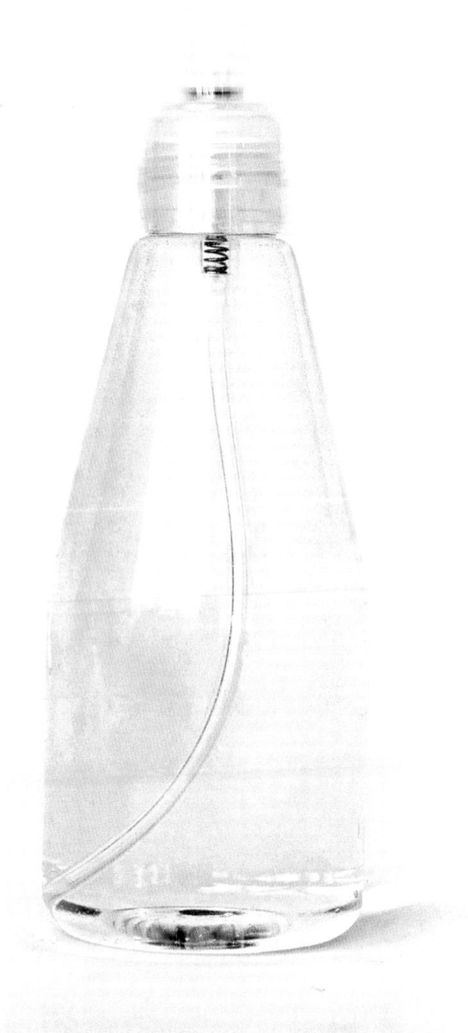

Add all the ingredients to a small spray bottle. Shake well and spray onto clothing and linens.

MOTH REPELLENT

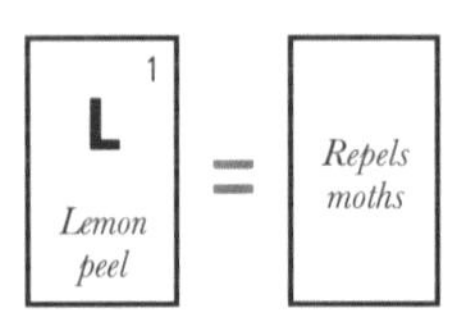

INGREDIENTS

1 lemon, freshly peeled
1 small muslin square, cut into 10 cm (4 in) squares
string

PREPARATION 5 minutes | **STORAGE TIME** they last for 2 weeks in drawer or wardrobe
STORAGE CONTAINER none

Put the lemon peel into a small square of muslin and lift up the edges to form a little sack. Tie the sack at the top with string to secure the ball of lemon peel, then tie around a hanger or put them in a drawer. You can reuse this muslin when you replace the peel.

ELECTRONIC SCREEN CLEANING SOLUTION

C_2H_5OH [1] *Alcohol*	+	H_2O [2] *Water*	=	*Cleans & evaporates*

INGREDIENTS
60 ml (2 fl oz/¼ cup) rubbing alcohol
120 ml (4 fl oz/½ cup) distilled water
6 small microfibre cloths

PREPARATION 5 minutes | **STORAGE TIME** up to 1 month
STORAGE CONTAINER airtight container, Kilner (Mason) jar, ziplock bag or plastic container

Mix the alcohol and distilled water together in a jug. Add the cloths, then squeeze each cloth to get rid of excess solution. Roll and put into your container. Before using, wring them out. They should barely be damp. Wipe the electronic screen from top to bottom and left to right.

CURTAIN SPRAY

EO [1] *Essential oils*	+	H_2O [2] *Water*	=	*Freshens*

INGREDIENTS

5 drops bergamot essential oil
5 drops lavender essential oil
5 drops clary sage essential oil
50 ml (1¾ fl oz/3 tablespoons) distilled water

PREPARATION 5 minutes | **STORAGE TIME** up to 1 month | **STORAGE CONTAINER** small glass or plastic spray bottle

Add all the ingredients to your spray bottle and shake to mix.
Spray liberally on to curtains.

ALTERNATIVE INSECT REPELLENT USING DRIED HERBS

CH_3COOH [1] Vinegar	+	H [2] Dried herbs	=	Repels insects

INGREDIENTS

1 litre (34 fl oz/4 cups) apple cider vinegar
2 tablespoons dried sage
2 tablespoons dried rosemary
2 tablespoons dried lavender
2 tablespoons dried thyme
2 tablespoons dried mint

PREPARATION 14 days | **STORAGE TIME** up to 3 months in refrigerator
STORAGE CONTAINER 1 litre (34 fl oz/4 cup) glass container to make and small spray bottle to store

Put all the ingredients into a 1 litre (34 fl oz/4 cup) glass container. Close or cover it, then shake the mixture once a day for 14 days. Strain the solution to remove the herbs and pour into your spray bottle. Spray as and when needed. To use on skin, dilute with 1 litre (34 fl oz/4 cups) water.

ROOM SCENT SPRAY

EO[1] Essential oils	+	H_2O[2] Water	=	Freshens

INGREDIENTS

4 drops lemon essential oil
4 drops bergamot essential oil
2 drops sandalwood essential oil
3 drops ylang-ylang essential oil
75 ml (2½ fl oz/5 tablespoons) distilled water

PREPARATION 5 minutes | **STORAGE TIME** up to 3 weeks | **STORAGE CONTAINER** small glass or plastic spray bottle

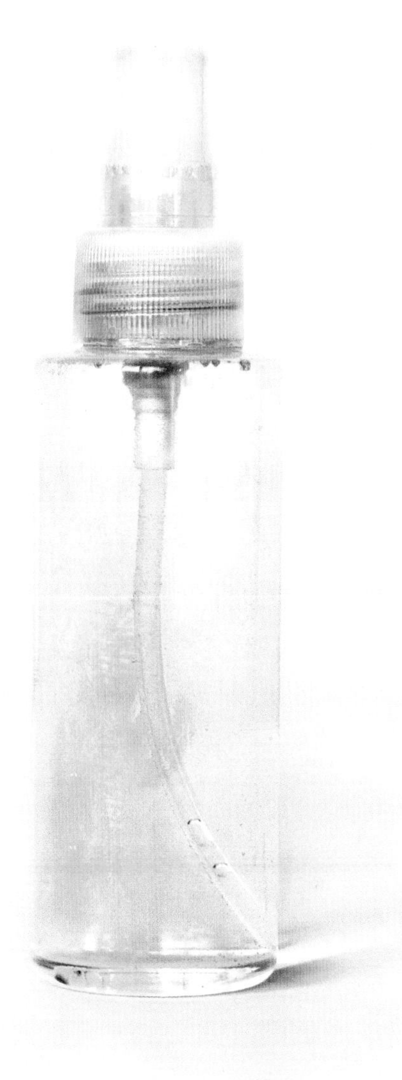

Add all the ingredients to your spray bottle. Spray the room directly, or spray the solution onto cotton wool balls and place these around the room.

SMOKE DEODORISER

EO [1] *Essential oils*	+	H_2O [2] *Water*	=	*Cleans the air*

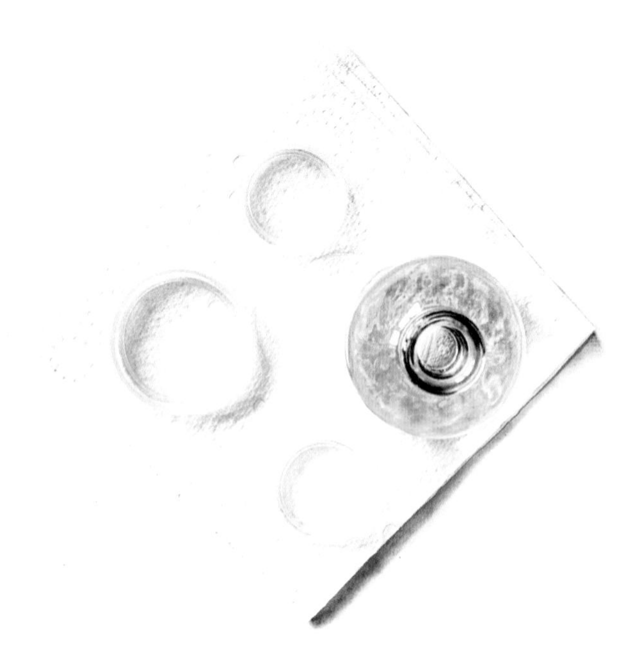

INGREDIENTS

4 drops tea tree essential oil
4 drops rosemary essential oil
4 drops eucalyptus essential oil
50 ml (1¾ fl oz/3 tablespoons) distilled water

PREPARATION 5 minutes | **STORAGE TIME** up to 2 weeks | **STORAGE CONTAINER** small glass or plastic spray bottle

Add all the ingredients to your spray bottle and spray where necessary around the house.

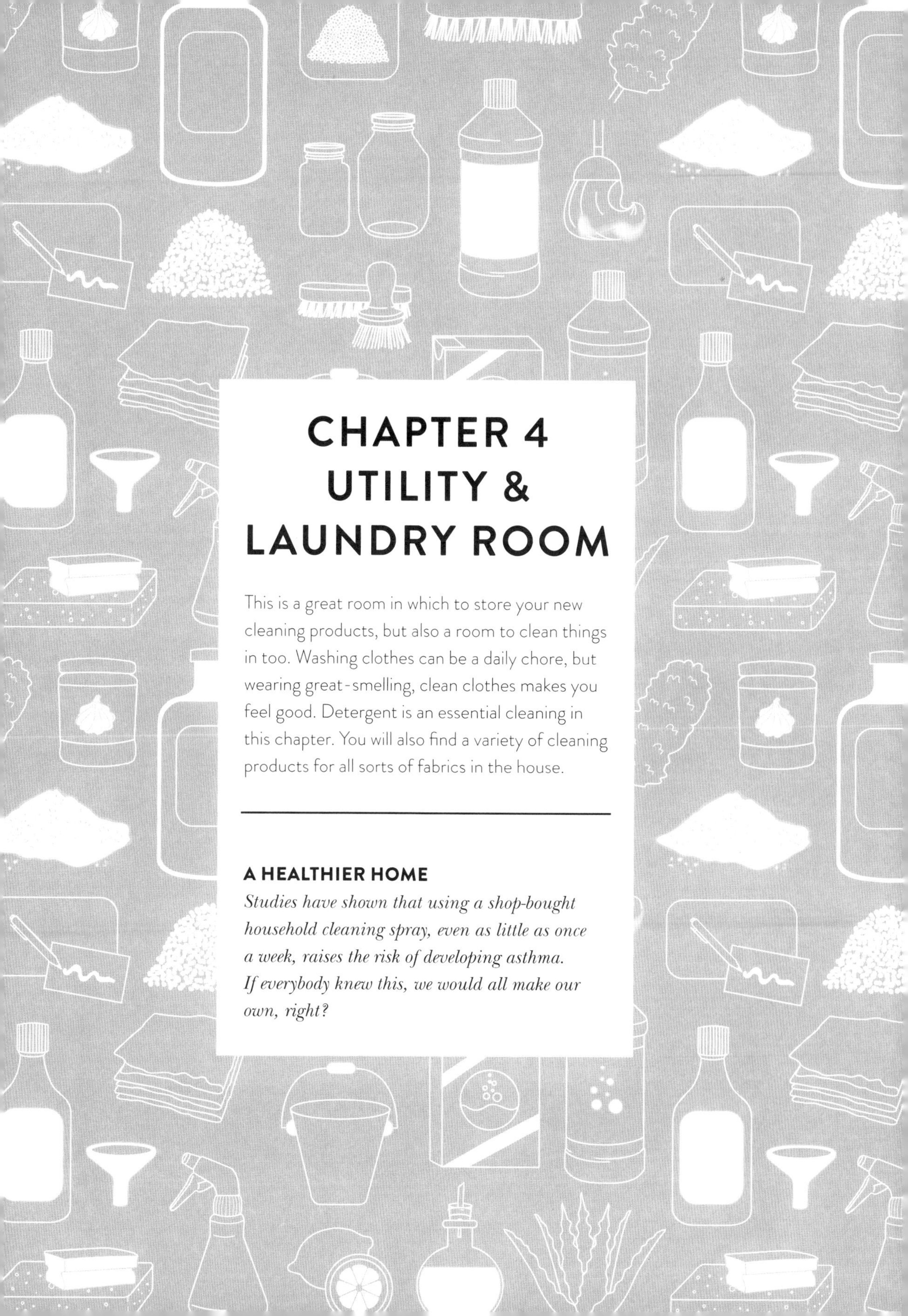

CHAPTER 4 UTILITY & LAUNDRY ROOM

This is a great room in which to store your new cleaning products, but also a room to clean things in too. Washing clothes can be a daily chore, but wearing great-smelling, clean clothes makes you feel good. Detergent is an essential cleaning in this chapter. You will also find a variety of cleaning products for all sorts of fabrics in the house.

A HEALTHIER HOME

Studies have shown that using a shop-bought household cleaning spray, even as little as once a week, raises the risk of developing asthma. If everybody knew this, we would all make our own, right?

WASHING DETERGENT (POWDER)

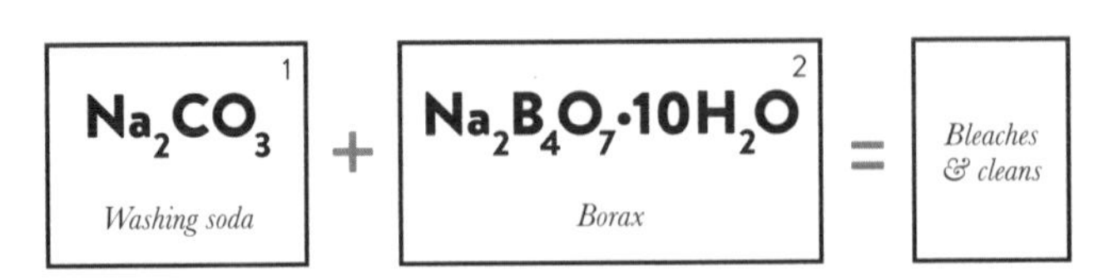

INGREDIENTS

400 ml (13 fl oz/generous 1½ cups) washing soda or sodium carbonate
400 g (14 oz) borax
200 g (7 oz) Castile soap bar, grated
2 teaspoons bicarbonate of soda (baking soda)

PREPARATION 5–10 minutes | **STORAGE TIME** up to 1 month | **STORAGE CONTAINER** large glass jar with lid

Put the washing soda, borax and grated soap into a large jar. Sprinkle the bicarbonate of soda over the mixture, cover with the lid and shake to mix. Use about 60 ml (2 fl oz/¼ cup) for each wash.

WASHING DETERGENT (LIQUID)

Na_2CO_3 [1]	+	$Na_2B_4O_7 \cdot 10H_2O$ [2]	=	*Softens water & cleans stains*
Washing soda		*Borax*		

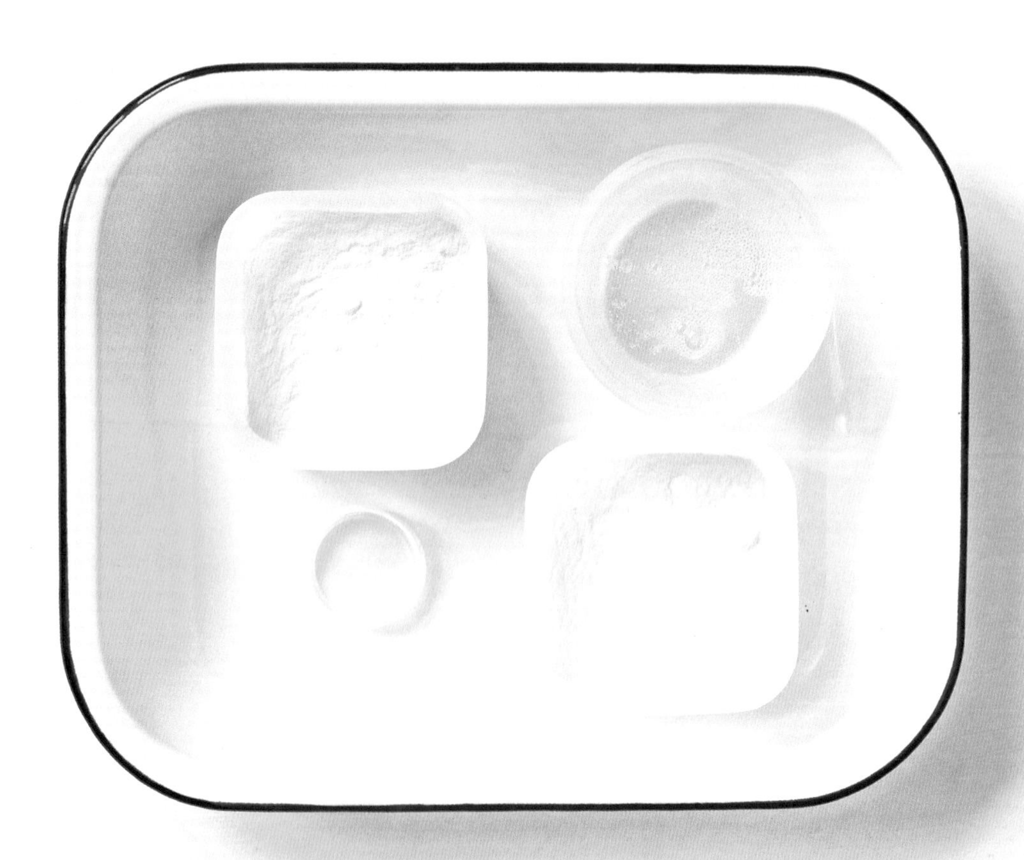

INGREDIENTS

250 g (9 oz) borax
250 g (9oz) washing soda or sodium carbonate
250 ml (8½ fl oz/1 cup) Castile liquid soap
15 drops lemon essential oil

PREPARATION 15–20 minutes | **STORAGE TIME** up to 1 month | **STORAGE CONTAINER** glass jars

Boil 1.5 litres (50 fl oz/6 cups) water, then turn off the heat and add the borax and washing soda. Stir thoroughly. In a bucket, mix the Castile soap with 2.5 litres (85 fl oz/10 cups) room temperature water. Add the lemon essential oil. Pour the hot mixture into the bucket and stir. Pour into glass jars for storage. Use 60 ml (2 fl oz/¼ cup) for each wash. If you have a stubborn stain, put some detergent directly on the stain and leave for 30 minutes before washing with the rest of your load.

LAUNDRY TABLETS

Na_2CO_3 [1]	+	S [2]	+	NaCl [3]	+	EO [4]	=	*Removes stains & freshens*
Washing soda		*Soap*		*Salt*		*Essential oils*		

INGREDIENTS

75 g (2½ oz) Castile soap bar, grated
300 g (10½ oz) washing soda or sodium carbonate
2 tablespoons Epsom salts
2 tablespoons hydrogen peroxide
10 drops lemon essential oil
10 drops thyme essential oil (you can use another combination of essential oils, if you like)
2½ tablespoons vinegar

PREPARATION 10 minutes, plus overnight | **STORAGE TIME** up to 3 months
STORAGE CONTAINER glass jar or ziplock bag

Grate the Castile soap into a bowl, add the washing soda and Epsom salts. Add the hydrogen peroxide slowly and stir. Add the essential oils and mix them in well. The mix should have the texture of wet sand.

Using an ice-cream scoop, scoop the mixture on to parchment paper, or use a half-round silicone mould. Spray with a mix of the vinegar and water in equal parts and leave overnight, by which time they should be hard to touch. Some might need a little longer. Store in a jar or plastic bag until ready for use.

FABRIC SOFTENER

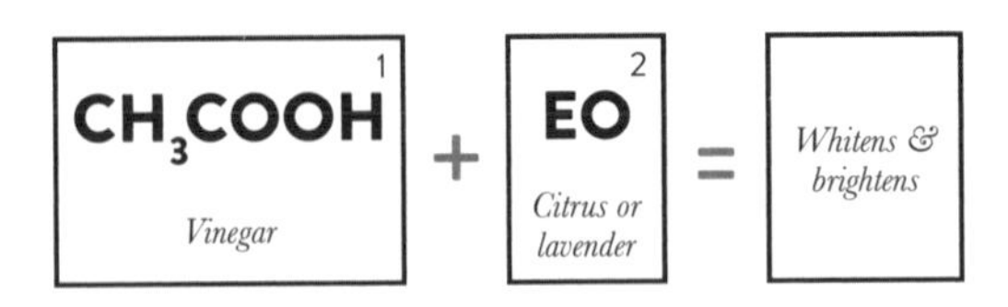

INGREDIENTS

1 litre (34 fl oz/4 cups) white vinegar

10 drops citrus or lavender essential oil (a fresh fragrance, or a more intense one)

PREPARATION 5 minutes | **STORAGE TIME** up to 3 months
STORAGE CONTAINER 1 litre (34 fl oz/4 cup) glass jar or bottle

Pour the vinegar into the glass container and add the drops of either fresh essential oil. Shake well before use. Add 125 ml (4 fl oz/½ cup) to the softener dispenser for a medium load and 250 ml (8½ fl oz/1 cup) for a large load.

WET WIPES

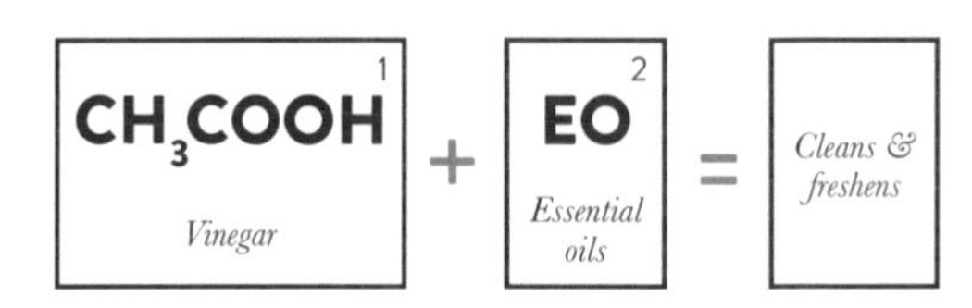

INGREDIENTS

250 ml (8½ fl oz/1 cup) distilled water
200 ml (7 fl oz/scant 1 cup) white vinegar
15 drops lemon essential oil
10 drops lavender essential oil
5 drops bergamot essential oil
about 12 cotton cloths, cut in half

PREPARATION 5–10 minutes | **STORAGE TIME** up to 1 week
STORAGE CONTAINER 1 litre (34 fl oz/4 cup) Kilner (Mason) jar

Fill the glass jar halfway with the distilled water. Add the vinegar and essential oils and stir well. Layer up your cloths into the jar and, once full, add more distilled water to top up the jar. Seal and shake well to distribute the oils evenly. Take a wipe, use it and return it to the jar afterwards. After substantial use, rinse with clean water and return to the jar.

SHOE ODOUR REPELLENT

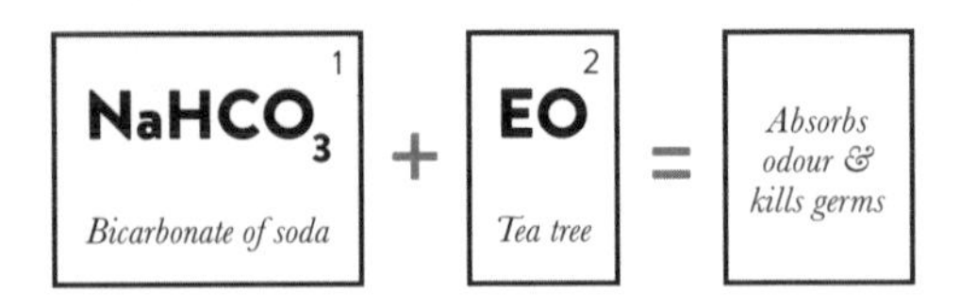

INGREDIENTS

1 tablespoon bicarbonate of soda (baking soda)

2–3 drops tea tree essential oil

PREPARATION 5 minutes | **STORAGE TIME** one-time product, no need for storage | **STORAGE CONTAINER** none

Sprinkle the bicarbonate of soda inside the shoes and add the essential oil. Shake the shoes to distribute the mixture, then leave for 10 minutes before pouring the mixture out or vacuuming. Repeat if needed.

WINE STAIN REMOVER

$NaCl$ [1]	+	$NaHCO_3$ [2]	+	$Na_2B_4O_7 \cdot 10H_2O$ [3]	=	*Gets rid of stains*
Salt		*Bicarbonate of soda*		*Borax*		

INGREDIENTS

30 g (1 oz) coarse sea salt
30 g (1 oz) bicarbonate of soda (baking soda)
30 g (1 oz) borax

PREPARATION 5 minutes | **STORAGE TIME** one-time product, no need for storage | **STORAGE CONTAINER** none

Mix all the ingredients together. Sprinkle over the wine stain and leave for 10–15 minutes. Vacuum and repeat if necessary. This mix can also be used on clothes; just use a brush to remove the powder.

WRINKLE RELEASER

CH_3COOH [1] *Vinegar*	+	C_2H_5OH [2] *Alcohol*	+	EO [3] *Lavender*	=	*Softens creases & freshens*

INGREDIENTS

1 teaspoon white vinegar
1 teaspoon vodka
2 drops lavender essential oil
250 ml (8½ fl oz/1 cup) distilled water

PREPARATION 5 minutes | **STORAGE TIME** up to 3 months | **STORAGE CONTAINER** glass or plastic spray bottle

Combine the ingredients together and pour into a spray bottle. Shake to combine. Spray on hanging clothing using a very light mist setting to moisten. Smooth the fabric, then let it air-dry and the wrinkles will disappear.

WASHING MACHINE CLEANER

INGREDIENTS

1 litre (34 fl oz/4 cups) white vinegar

PREPARATION 1 hour, plus washing cycle | **STORAGE TIME** one-time product, no need for storage
STORAGE CONTAINER none

Choose the longest cycle, the hottest water and the largest load setting on your washing machine, and add the vinegar. Start the machine. When the drum is full, pause the cycle for 1 hour, then let the cycle finish. Leave the door open to let the interior dry.

CLOTHES WHITENER

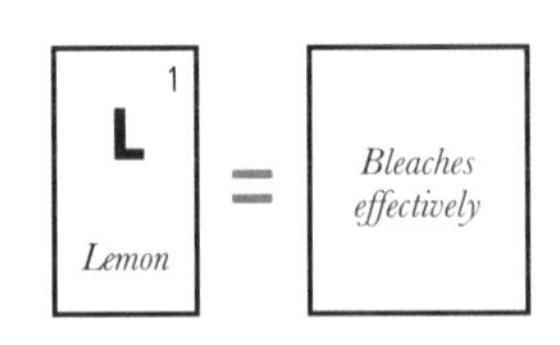

INGREDIENTS

120 ml (4 fl oz/½ cup) lemon juice

PREPARATION 5 minutes | **STORAGE TIME** one-time product, no need for storage | **STORAGE CONTAINER** none

Mix the lemon juice with 5 litres (170 fl oz/21 cups) hot water in a bucket or laundry tub and leave your whites to soak for up to 2 hours. Run through a rinse cycle and set out to dry, preferably in the sun. Wash afterwards as usual.

LAUNDRY WHITENER

H_2O_2 [1]		Na_2CO_3 [2]		
Hydrogen peroxide	+	*Washing soda*	=	*Gets rid of stains*

INGREDIENTS

2 tablespoons hydrogen peroxide

1 tablespoon washing soda or sodium carbonate

PREPARATION 20 minutes | **STORAGE TIME** one-time product, no need for storage | **STORAGE CONTAINER** none

Mix the ingredients together in a small bowl and microwave on high for 15 seconds to dissolve the washing soda. The solution will fizz, so avoid contact with the eyes. Stir to help dissolve. Apply directly to the stain and leave for 15 minutes. Wash as usual.

STAIN SPRAY

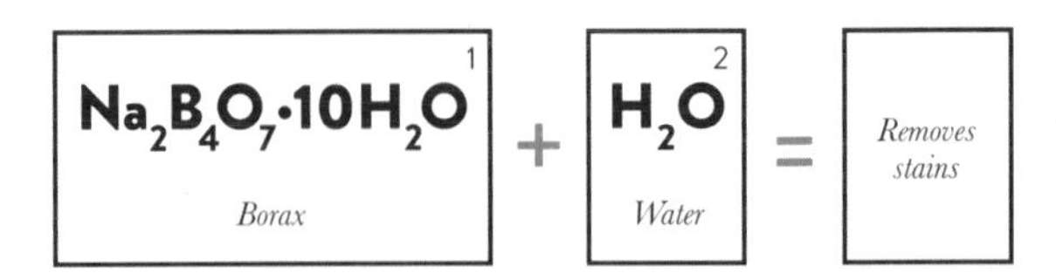

INGREDIENTS

2 tablespoons borax

230 ml (8 fl oz/scant 1 cup) warm distilled water

PREPARATION 5 minutes | **STORAGE TIME** up to 1 month | **STORAGE CONTAINER** small glass or plastic spray bottle

Mix the borax with the distilled water and pour into a spray bottle. Shake thoroughly and spray directly on the stain. Wash as usual. If you have a particularly greasy stain, try using chalk to pre-treat before washing.

TEA TREE STAIN STICK

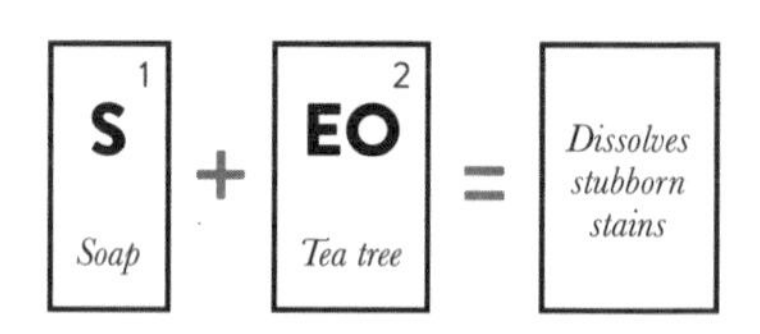

INGREDIENTS

75 g (2½ oz) Castile soap bar, grated
4 drops tea tree essential oil

PREPARATION 15 minutes | **STORAGE TIME** up to 6 months | **STORAGE CONTAINER** old deodorant stick container

Place the grated soap and 3 teaspoons water in a bowl. Microwave at 20-second intervals for 1 minute. Stir and check to see if it has melted. Keep going until the soap has melted. Leave to cool for 1 minute, then stir in the essential oil. Pour into a clean push-up container and allow to cool to a solid state. Use on any stains, then launder as usual.

OXYGEN BLEACH PASTE

H_2O_2 (1) *Hydrogen peroxide*	+	Na_2CO_3 (2) *Washing soda*	+	H_2O (3) *Water*	=	*Removes stubborn stains*

INGREDIENTS

1 tablespoon hydrogen peroxide

1 tablespoon washing soda or sodium carbonate

PREPARATION 5 minutes | **STORAGE TIME** once mixed, product will slowly lose effectiveness, so no storage necessary
STORAGE CONTAINER none

Mix the peroxide and washing soda in a small bowl into a dry paste. Add a little water to the bowl before putting it into the detergent dispenser in the washing machine. Wash as usual.

LINEN FRESHENER

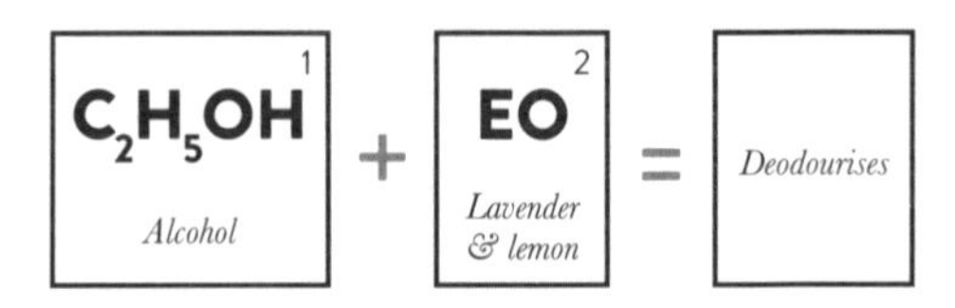

INGREDIENTS

120 ml (4 fl oz/½ cup) vodka
10 drops lavender essential oil
5 drops lemon essential oil

PREPARATION 5 minutes | **STORAGE TIME** up to 2 months | **STORAGE CONTAINER** glass or plastic spray bottle

Combine the vodka with 250 ml (8½ fl oz/1 cup) water. Decant into a small spray bottle and add the essential oils. Spray about 20 cm (8 in) away from the linen.

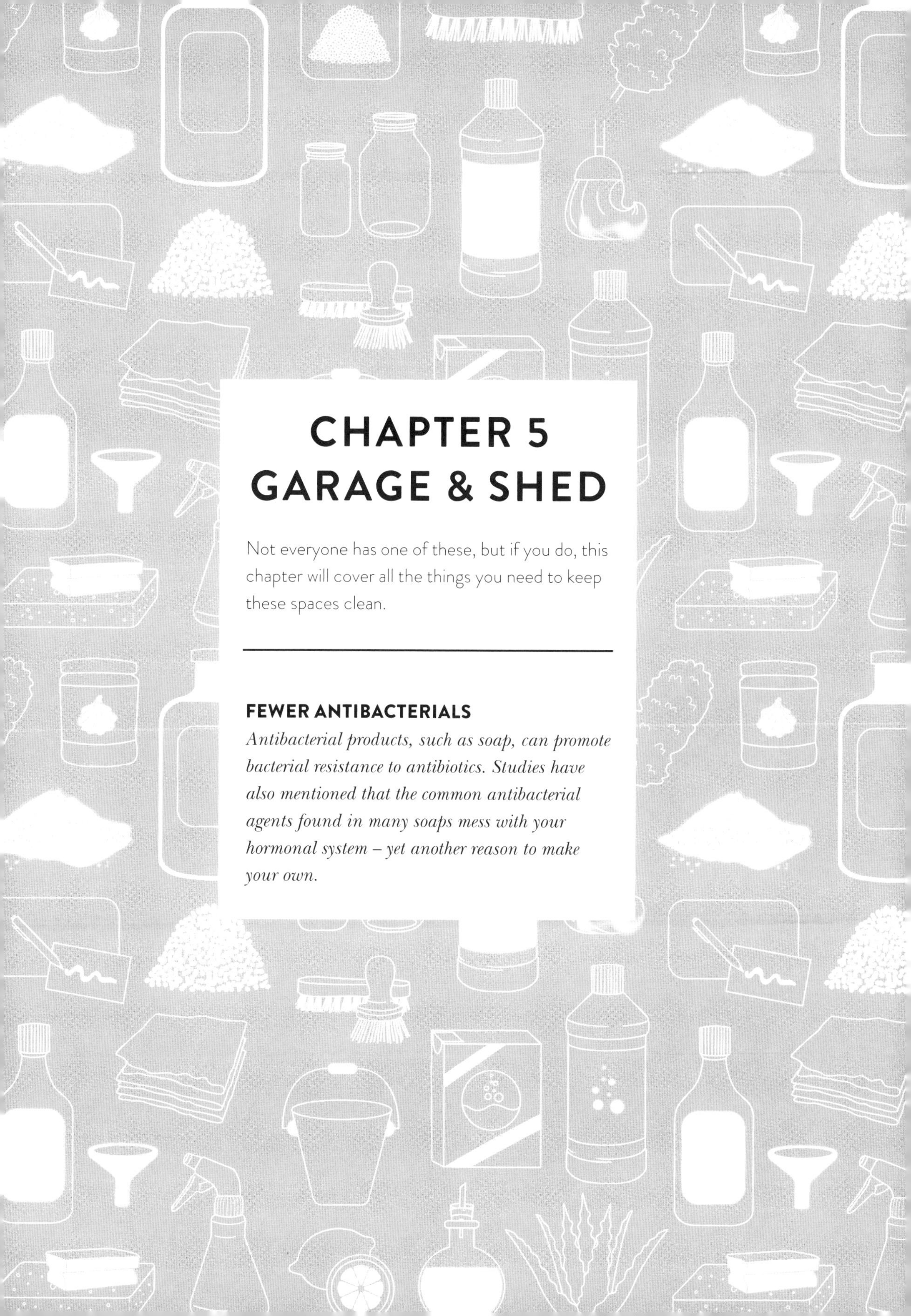

CHAPTER 5
GARAGE & SHED

Not everyone has one of these, but if you do, this chapter will cover all the things you need to keep these spaces clean.

FEWER ANTIBACTERIALS

Antibacterial products, such as soap, can promote bacterial resistance to antibiotics. Studies have also mentioned that the common antibacterial agents found in many soaps mess with your hormonal system – yet another reason to make your own.

PREPARATION 5 minutes | **STORAGE TIME** one-time product, no need for storage | **STORAGE CONTAINER** none

TERMITE REPELLENT

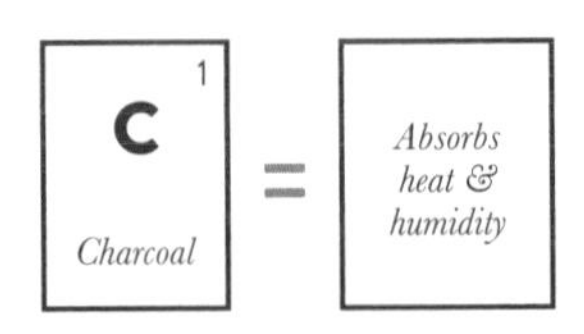

INGREDIENTS

1 binchotan charcoal stick

Place the binchotan stick near areas where the termites are entering or could enter. The stick absorbs the heat and humidity, which the termites like, so this will discourage them.

PREPARATION 5 minutes | **STORAGE TIME** one-time product, no need for storage | **STORAGE CONTAINER** none

ANT REPELLENT

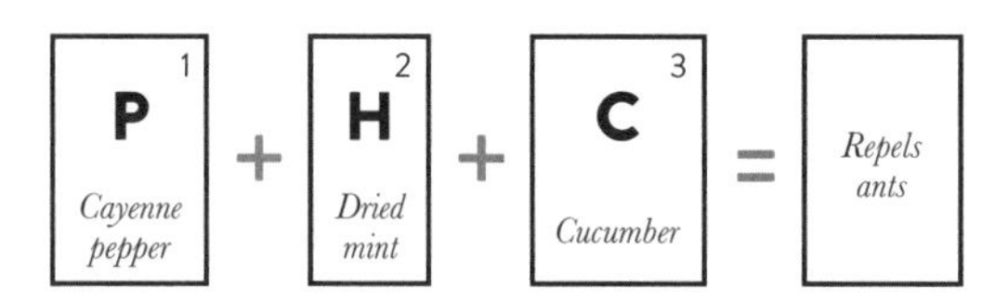

INGREDIENTS

cayenne pepper
dried mint
cloves
fresh cucumber slices

Sprinkle cayenne pepper across all possible entry points, especially where you see a line of ants entering. Dried mint, cloves and cucumber can be left on work surfaces to discourage ants from exploring further.

MOSQUITO & FLY REPELLENT

C_2H_5OH [1] *Alcohol*	+	A [2] *Almond oil*	+	EO [3] *Essential oils*	=	*Repels insects*

INGREDIENTS

2 tablespoons vodka
2 tablespoons sweet almond oil
50 drops eucalyptus essential oil
15 drops lavender essential oil
15 drops rosemary essential oil

PREPARATION 10 minutes | **STORAGE TIME** up to 2 months

STORAGE CONTAINER small glass or plastic spray bottle

Add all the ingredients to a spray bottle and shake it to combine. Spray into a room or directly onto your skin for protection.

TICK & FLEA REPELLENT FOR PETS

H_2O [1] *Water*	+	EO [2] *Essential oils*	=	*Repels insects*

INGREDIENTS

100 ml (3½ fl oz/scant ½ cup) distilled water
5 drops lavender essential oil
5 drops tea tree essential oil
5 drops geranium essential oil

PREPARATION 5–10 minutes | **STORAGE TIME** up to 2 months
STORAGE CONTAINER small plastic or glass spray bottle

Pour the distilled water into the spray bottle and add the essential oils. Shake. Before using, test a small area of your pet's skin for sensitivity. Lightly spray the mixture on your pet's fur to repel ticks and fleas. Repeat if the problem persists.

COCKROACH REPELLENT

$Na_2B_4O_7 \cdot 10H_2O$ [1]	+	$NaHCO_3$ [2]	+	$MgSO_4$ [3]	=	*Kills bugs*
Borax		*Bicarbonate of soda*		*Epsom salts*		

INGREDIENTS

100 g (3½ oz) borax
100 g (3½ oz) bicarbonate of soda (baking soda)
100 g (3½ oz) Epsom salts

PREPARATION 5 minutes | **STORAGE TIME** one-time product, no need for storage | **STORAGE CONTAINER** none

Mix all the ingredients together in a small bowl. Leave on surfaces or in any cupboards you have seen a cockroach.

PET FOOT WIPES

S [1] *Soap*	+	EO [2] *Clove & lemon*	=	*Odour-reducing cleaner*

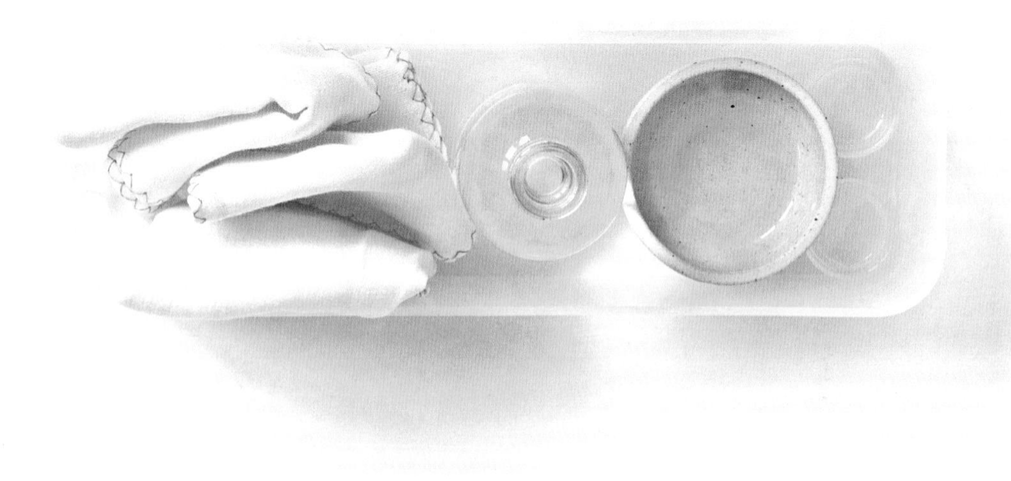

INGREDIENTS

12 cotton washcloths
1 tablespoon Castile liquid soap
2 drops clove essential oil
2 drops lemon essential oil
360 ml (12 fl oz/1½ cups) distilled water

PREPARATION 10 minutes | **STORAGE TIME** up to 1 week
STORAGE CONTAINER airtight container, Kilner (Mason) jar or ziplock bag

Put the washcloths into your airtight container. Mix the rest of the ingredients in a measuring jug and pour over the cloths. Seal the container and shake until as much liquid as possible has been absorbed. Pour out any excess liquid. Remove a cloth and squeeze out any remaining liquid, then wipe the paws of your pet. Washcloths can be laundered and reused.

GARDEN TOOL CLEANING KIT

S [1] *Soap*	+	**EO** [2] *Rosemary*	=	*Cleans & shines*

INGREDIENTS

1 tablespoon Castile liquid soap
5 drops rosemary essential oil

PREPARATION 5 minutes | **STORAGE TIME** one-time product, no need for storage | **STORAGE CONTAINER** none

In a large bucket big enough to fit all the tools, combine the soap with 5–10 litres (170–340 fl oz/21–42 cups) warm water, enough to cover the tools. Soak for a few minutes, then scrub with a brush. Dry thoroughly. Use the rosemary oil to oil the tools.

PATIO CLEANER

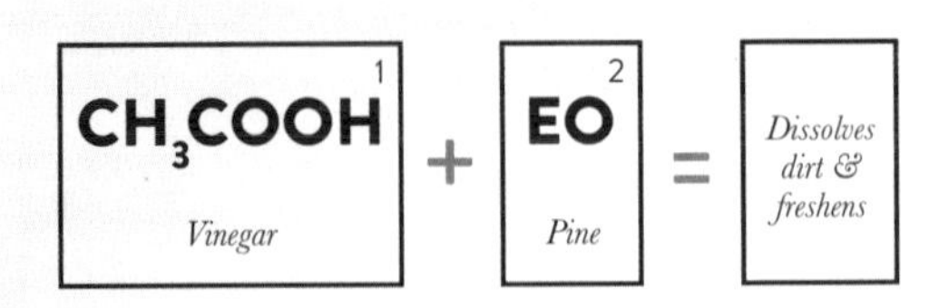

INGREDIENTS

500 ml (17 fl oz/2 cups) white vinegar
20 drops pine essential oil
250 ml (8½ fl oz/1 cup) white vinegar (optional)
100 g (3½ oz) coarse salt (optional)

PREPARATION 5 minutes | **STORAGE TIME** one-time product, no need for storage | **STORAGE CONTAINER** none

In a 1 litre (34 fl oz/4 cup) plastic or glass spray bottle, add the 500 ml (17 fl oz/2 cups) vinegar and top up with water. Add the essential oil and shake. Spray over the concrete patio and wait for 20 minutes before mopping up. If you have some tough stains, mix the 250 ml (8½ fl oz/1 cup) vinegar with the coarse salt, then use a brush to scrub the area. Please note that the vinegar will also kill the weeds in between the concrete slabs.

GARDEN FURNITURE CLEANER

S[1]	+	EO[2]	+	H_2O[3]	=	
Soap		*Bergamot*		*Water*		*Cleans & shines*

INGREDIENTS

2 tablespoons Castile liquid soap

2 drops bergamot essential oil

PREPARATION 5 minutes | **STORAGE TIME** indefinite | **STORAGE CONTAINER** glass or plastic spray bottle

Add the liquid soap and 450 ml (15½ fl oz/1¾ cups) warm water to a spray bottle. Top with the essential oil and shake to combine. Wet furniture using a hose, then spray the solution, rinse and air-dry.

PAINTBRUSH CLEANER

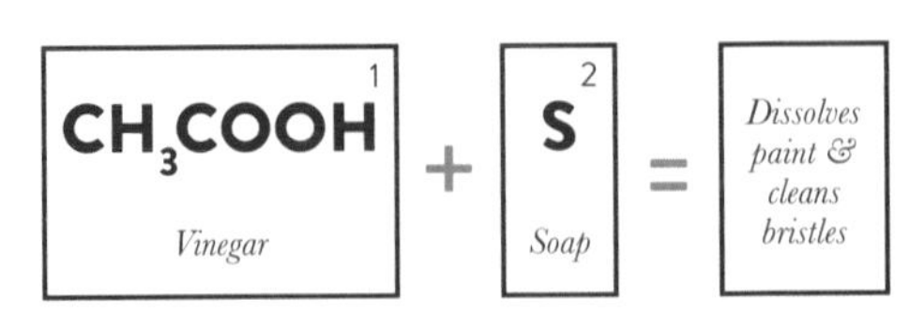

INGREDIENTS

250 ml (8½ fl oz/1 cup) white vinegar
1 tablespoon Castile liquid soap

PREPARATION 1 hour | **STORAGE TIME** one-time product, no need for storage | **STORAGE CONTAINER** none

Heat the vinegar in a heatproof bowl in the microwave for 2 minutes. Place the paintbrush (bristles only) in the vinegar for 1 hour, or overnight. Rinse and repeat if necessary. Apply the soap and wash thoroughly. Dry excess water with a clean cloth or paper towels and air-dry.

CAR WASHING SOLUTION

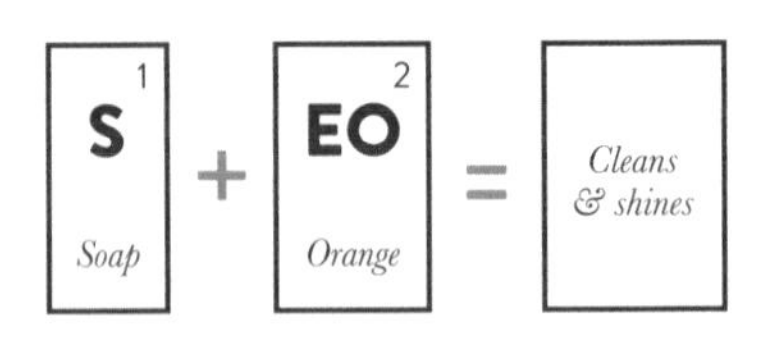

INGREDIENTS

2 tablespoons Castile liquid soap
3 drops orange essential oil
microfibre cloth

PREPARATION 5 minutes | **STORAGE TIME** one-time product, no need for storage | **STORAGE CONTAINER** none

Add the soap, essential oil and cloth to a large bucket, then fill it up with hot water. Wet the car with a hose, dip the cloth in the solution and start washing. Rinse and dry thoroughly.

INTERIOR CAR WIPES

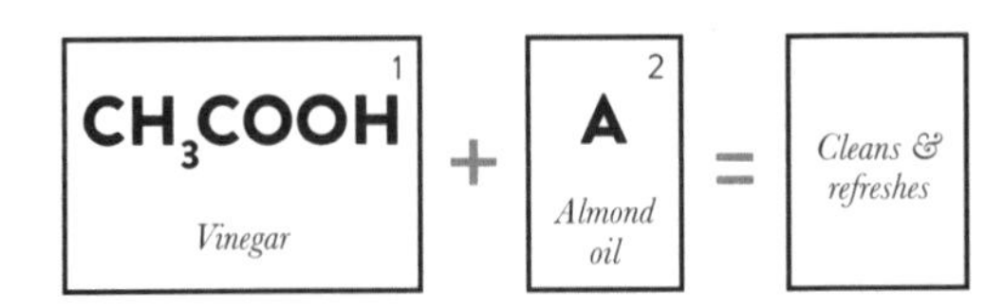

INGREDIENTS

12 sheets strong paper towel
120 ml (4 fl oz/½ cup) white vinegar
180 ml (6 fl oz/¾ cup) distilled water
3 tablespoons sweet almond oil
2 drops peppermint essential oil

PREPARATION 15 minutes | **STORAGE TIME** up to 1 week

STORAGE CONTAINER airtight container, 500 ml (17 fl oz/2 cup) Kilner (Mason) jar or ziplock bag

Cut the paper towels in half and put them in the airtight container. Mix the remaining ingredients together in a large jug and pour over the paper towels. Cover with a lid, then shake and swirl the solution, making sure the towels have absorbed all the liquid. Pour off any excess; the wipe works better if on the dry side. When ready to clean, carefully remove a wipe and wring out excess before using.

BBQ GRILL CLEANER

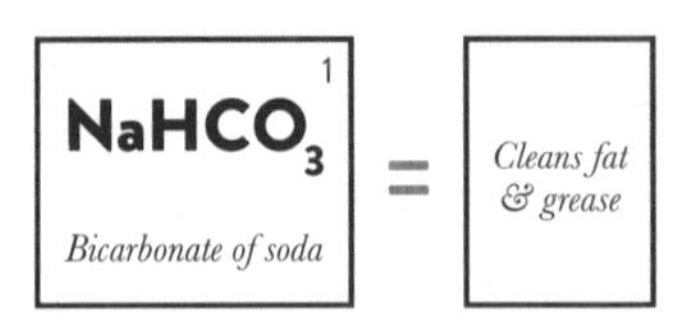

INGREDIENTS

120 g (4 oz) bicarbonate of soda (baking soda)
250 ml (8½ fl oz/1 cup) white vinegar

PREPARATION 5 minutes | **STORAGE TIME** one-time product, no need for storage | **STORAGE CONTAINER** none

Make a paste with the bicarbonate of soda and 240 ml (8 fl oz/scant 1 cup) warm water in a bowl. Apply the paste to the grill and leave for 15 minutes. Meanwhile, using the vinegar, clean the exterior of the barbecue using paper towels or a cloth. No rinsing necessary. After cleaning, remove the grill and scrub in the sink or outside with a brush. Rinse thoroughly and allow to air-dry.

YOGA MAT CLEANER

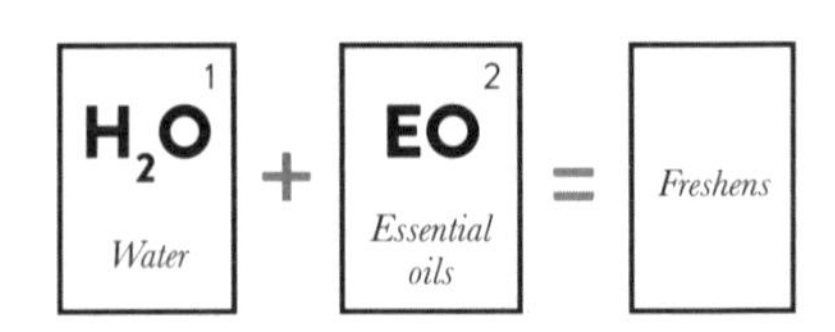

INGREDIENTS

2 tablespoons distilled water
10 drops lavender essential oil
4 drops tea tree essential oil
5 drops lemongrass essential oil

PREPARATION 5 minutes | **STORAGE TIME** up to 2 months
STORAGE CONTAINER small glass or plastic spray bottle

Pour the distilled water into a spray bottle and add the essential oils.
Shake and spray on to your yoga mat section by section. Rub it in using a clean cloth.

EXERCISE MACHINE SPRAY

H_2O [1] *Water*	+	EO [2] *Lemon & tea tree*	=	*Disinfects & freshens*

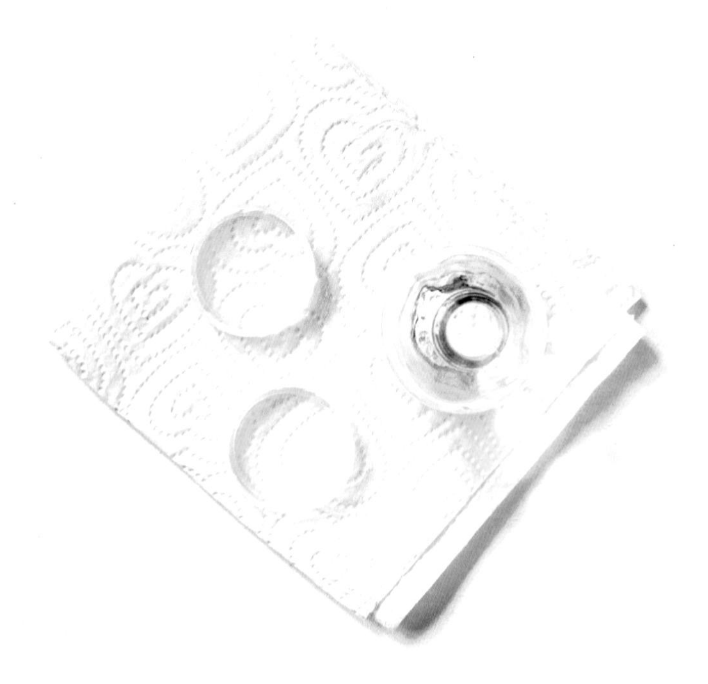

INGREDIENTS

2 tablespoons distilled water
10 drops lemon essential oil
10 drops tea tree essential oil

PREPARATION 5 minutes | **STORAGE TIME** up to 2 weeks | **STORAGE CONTAINER** small glass or plastic spray bottle

Add all the ingredients to a spray bottle and shake.
Spray on exercise equipment and wipe with a fresh cloth.

TRAINER ODOUR REMOVER

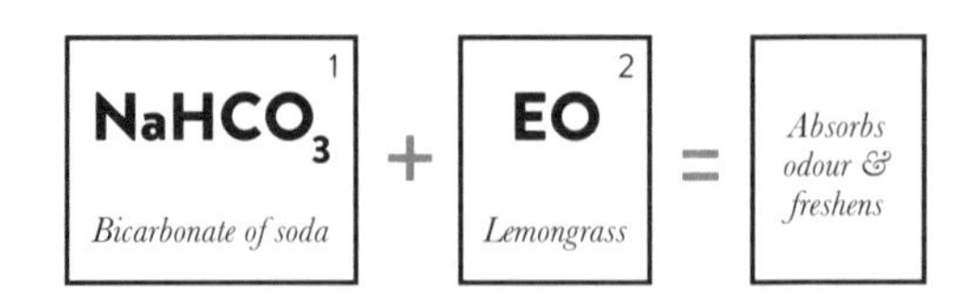

INGREDIENTS

2 tablespoons bicarbonate of soda (baking soda)
5 drops lemongrass essential oil

PREPARATION 5 minutes | **STORAGE TIME** one-time product, no need for storage | **STORAGE CONTAINER** none

In a small bowl, mix the ingredients together. Sprinkle evenly in the heel of the trainer, then tilt and let the powder work downwards towards the toe. Leave for up to 24 hours, then shake out all the loose powder.

INDEX

Entries in italics refer to recipe names.

A

Air Freshener 130
Air Purifier 132
Air Vent Cleaner 54
alcohol 9, 46, 108, 150, 164, 180
All-purpose Bathroom Cleaner 20
All-purpose Wipes 166
All-purpose Work Surface Spray 66
almond oil 9, 156, 158, 162, 226, 244
aloe vera gel 9, 62
Alternative Insect Repellent Using Dried Herbs 184
Antibacterial Handwash 96
Antibacterial Wipes 170
Ant Repellent 225

B

Bathroom Air Freshener 40
Bathroom Disinfectant Spray 44
Bathroom Liquid Hand Soap 58
BBQ Grill Cleaner 246
Bed Bug Repellent 168
bergamot essential oil 158, 182, 186, 200, 238
bicarbonate of soda 8, 26, 28, 30, 32, 34, 38, 48, 56, 78, 80, 82, 84, 88, 102, 116, 118, 126, 134, 136, 192, 202, 204, 230, 246, 252
binchotan charcoal stick 70, 124, 132, 133
borax 8, 74, 78, 134, 140, 152, 192, 194, 204, 214, 224, 230

C

Carpet Cleaner (Powder) 134
Carpet Stain Remover 136
Car Washing Solution 242
Castile soap 8, 20, 38, 44, 58, 62, 66, 74, 76, 80, 92, 94, 106, 108, 112, 115, 116, 122, , 172, 192, 194, 196, 216, 232, 234, 238, 240, 242
cayenne pepper 225
cedarwood essential oil 174
Chopping Board Cleaner 90
citric acid 28, 30, 74
citronella essential oil 11
Citrus & Lavender Toilet Fizz Tablets 28
Citrus & Rosemary Liquid Hand Soap 92
Citrus Foaming Hand Soap 94
clary sage essential oil 182
cleaning kit 16
Clothes Whitener 210
clove essential oil 11, 102, 150, 176, 232
Cockroach Repellent 230
coconut oil 9, 68, 126
Coffee Pot Cleaner 114
cornflour 46, 64
cucumber 225
Curtain Spray 182

D/E

Daily Bath & Shower Cleaner 36
Daily Sink Scrub 102
Daily Toilet Cleaner 24
Damp Defier 133
Dish Soap 62
Dishwasher Detergent (Liquid) 76
Dishwasher Detergent (Powder) 74
Disposable Furniture Wipes 158
Door & Skirting Board Cleaner 140
Drain Unblocker 88
Electronic Screen Cleaning Solution 180
essential oils 10
eucalyptus essential oil 11, 20, 40, 50, 130, 148, 174, 188
Exercise Machine Spray 250

F

Fabric Softener 198
Floor Wipes 154
Fruit & Veg Wash 104
Furniture Polish 144

G

Garden Furniture Cleaner 238
Garden Tool Cleaning Kit 234
garlic 168
geranium essential oil 228, 168
Glass & Mirror Wipes 164
Granite Cleaner 108
grapefruit essential oil 11, 30, 62, 72
Grout Cleaner 32

H/I

Hand & Face Wipes 122
Hardwood Floor Cleaner 138
Heavy-duty Toilet Cleaner 26
herbs, dried 184, 225
hydrogen peroxide 9, 20, 34, 116, 196, 212, 218
Ice Machine Cleaner 115
Intense Bath & Shower Cleaner 38
Interior Car Wipes 244

K/L

Kettle Cleaner 98
Kids' Cleaning Wipes 120
Kitchen Bin Deodoriser 72
Knife Block Cleaner 112
Label or Sticky Residue Remover 126
Laminate Floor Cleaner 148
Laundry Tablets 196
Laundry Whitener 212
lavender essential oil 11, 28, 40, 58, 96, 120, 130, 134, 166, 174, 176, 182, 198, 200, 206, 220, 226, 228, 248
Leather Wipes 162

lemon 9, 32, 34, 64, 74, 76, 80, 86, 90, 98, 104, 110, 144, 156, 160, 178, 210
Lemon & Peppermint Tile Cleaner 48
lemon essential oil 11, 20, 28, 32, 36, 48, 54, 62, 76, 78, 92, 94, 102, 120, 144, 150, 152, 156, 166, 186, 194, 196, 200, 220, 232, 250
Lemon Dishwasher Detergent Tablets 78
lemongrass essential oil 248, 252
Lime & Grapefruit Toilet Fizz Tablets 30
lime essential oil 11, 30
Linen Freshener 220
Linen Spray 174

M/N

Marble Cleaner 106
Microwave Cleaner 86
Mildew Remover 22
Mirror & Glass Cleaner 46
Mosquito & Fly Repellent 226
Moth Repellent 178

O

olive oil 9, 122, 144, 146
orange 28, 156
Orange & Tea Tree Disinfectant Cleaner 100
orange essential oil 11, 28, 40, 54, 94, 100, 110, 130, 138, 146, 152, 154, 156, 158, 242
Oven & Hob Cleaner 82
Oven Cleaner 84
Oxygen Bleach Paste 218

P/Q

Paintbrush Cleaner 240
Patio Cleaner 236
peppermint essential oil 11, 20, 48, 58, 62, 72, 78, 134, 138, 164, 170, 244
Pet Foot Wipes 232
pine essential oil 236
Plastic Storage Container Cleaner 116
Quick Insect Repellent for Your Home 176

R

Refrigerator & Freezer Cleaner 80
Refrigerator Odour Deodoriser 70
Room Scent Spray 186
rosemary essential oil 11, 56, 72, 92, 176, 188, 226, 168, 234
Rust Remover 52

S

salt 9, 74, 84, 90, 196, 204, 230, 236
sandalwood essential oil 186
Scratch Remover 160
Shoe Odour Repellent 202
Shower Curtain Cleaner 50
Shower Head Cleaner 42
Silver Cleaner 118
Small Appliance Cleaner 110
Smoke Deodoriser 188
Stainless Steel Polish 68
Stain Spray 214
Strong Drain Cleaner 56

T

tea tree essential oil 11, 22, 24, 26, 40, 44, 66, 96, 100, 120, 130, 134, 154, 170, 188, 202, 216, 228, 168, 248, 250
Tea Tree Stain Stick 216
Termite Repellent 224
thyme essential oil 196
Tick & Flea Repellent for Pets 228
Tile Floor Cleaner 152
Trainer Odour Remover 252

U/V

Upholstered Fabric Cleaner 172
vegetable glycerine 62
vinegar 8, 22, 24, 26, 32, 36, 38, 42, 46, 48, 50, 52, 54, 56, 64, 68, 74, 78, 82, 86, 88, 100, 104, 110, 114, 120, 136, 138, 142, 146, 148, 150, 154, 156, 158, 160, 162, 164, 166, 170, 184, 196, 198, 200, 206, 208, 236, 240, 244, 246
Vinyl Floor Cleaning Spray 150
vitamin E oil 9, 58, 92, 94, 96
vodka 9, 96, 170, 206, 220, 226

W/Y

Washing Detergent (Liquid) 194
Washing Detergent (Powder) 192
Washing Machine Cleaner 208
washing soda 8, 62, 74, 100, 192, 194, 196, 212, 218
Water Purifier 124
Wall Scrub 142
Wet Wipes 200
Whitening Grout Cleaner 34
Window Cleaner 64
Wine Stain Remover 204
witch hazel 122, 176
Wood-cleaning Cloths 156
Wood Polish 146
Wrinkle Releaser 206
ylang-ylang essential oil 186
Yoga Mat Cleaner 248

Acknowledgements

A big thank you to Deidre Rooney for her fab photos in this book, and also a big thanks to Kathy Steer and Michelle Tilly for their tenacity and patience. This book wouldn't have happened if it wasn't for the help of my husband Jono. I love you always.

The author has researched each product used in this book but is not responsible for any adverse effects any of the products may have on an individual.

First published by Hachette Livre (Marabout) in 2018
This English language edition published in 2019 by Hardie Grant Books, an imprint of Hardie Grant Publishing

Hardie Grant Books (London)
5th & 6th Floors, 52–54 Southwark Street
London SE1 1UN

Hardie Grant Books (Melbourne)
Building 1, 658 Church Street
Richmond, Victoria 3121

hardiegrantbooks.com

British Library Cataloguing-in-Publication Data. A catalogue record for this book is available from the British Library.

Natural Home Cleaning by Fern Green
ISBN: 978-1-78488-239-6
10 9 8 7 6 5 4 3 2

For the French edition:

Publisher: Catie Ziller
Designer: Michelle Tilly
Photographer: Deirdre Rooney
Editor: Kathy Steer

For the English edition:

Publishing Director: Kate Pollard
Junior Editor: Rebecca Fitzsimons
Cover Design: Rebecca Fitzsimons
Editor: Kay Delves

Colour Reproduction by p2d
Printed and bound in China by 1010